T5-CUO-301

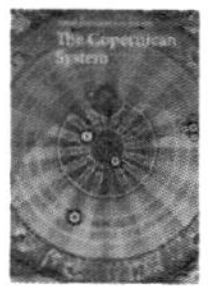

Copernican System **$31.95**

When Nicolaus Copernicus stated that Earth revolves around the sun, he initiated a centuries-long conflict between proponents of heliocentrism and those that maintained a geocentric view of the univers...

#2020592 E. Richardson Available:08/15/2017 128 pgs
Grade:9Y Dewey:909.82

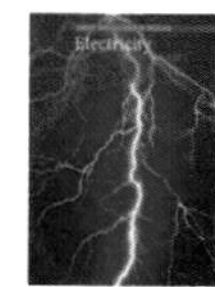

Electricity **$31.95**

Over time, Benjamin Franklin's kite and key experiment has taken on mythic proportions. Franklin's experiment established the relationship between lightning and electricity, but it would take the resea...

#2020590 P. Sherman Available:08/15/2017 128 pgs
Grade:9Y Dewey:537

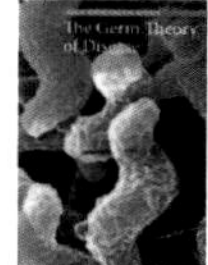

Germ Theory of Disease **$31.95**

From ancient times until the early nineteenth century, many medical practitioners believed that the body contained four humors: blood, yellow bile, black bile, and phlegm. Humoral doctrine stated that ...

#2020591 K. Thiel Available:08/15/2017 128 pgs
Grade:9Y Dewey:616

Semiconductors **$31.95**

The discovery of electricity fundamentally changed day-to-day life. Yet after electricity's discovery, scientists worked to find the best way to harness electrical currents. Today, semiconductors are k...

#2020634 G. Murphy Available:08/15/2017 128 pgs
Grade:9Y Dewey:537.62

Vaccination **$31.95**

Contemporary vaccination is rooted in centuries of scientific discovery. Some scholars believe that as far back as 1000 CE, Chinese Taoists used variolation (or inoculation) to control the spread of di...

#2020594 E. Richardson Available:08/15/2017 128 pgs
Grade:9Y Dewey:614.47

GREAT DISCOVERIES IN SCIENCE

Genetically Modified Crops

Megan Mitchell

Published in 2019 by Cavendish Square Publishing, LLC
243 5th Avenue, Suite 136, New York, NY 10016

First Edition

Library of Congress Cataloging-in-Publication Data

Names: Mitchell, Megan, author.
Title: Genetically modified crops / Megan Mitchell.
Description: First edition. | New York : Cavendish Square, 2019. | Series: Great discoveries in science | Includes bibliographical references and index.
Identifiers: LCCN 2018009543 (print) | LCCN 2018011493 (ebook) | ISBN 9781502643681 (ebook) | ISBN 9781502643780 (library bound) | ISBN 9781502643902 (pbk.)
Subjects: LCSH: Transgenic plants.
Classification: LCC SB123.57 (ebook) | LCC SB123.57 .M58 2019 (print) | DDC 631.5/233--dc23
LC record available at https://lccn.loc.gov/2018009543

Editorial Director: David McNamara
Editor: Jodyanne Benson
Copy Editor: Michele Suchomel-Casey
Associate Art Director: Alan Sliwinski
Designer: Christina Shults
Production Coordinator: Karol Szymczuk
Photo Research: J8 Media

The photographs in this book are used by permission and through the courtesy of:
Cover Lindsay Eyink/Wikimedia Commons/File:Hybrid corn Yellow Springs, Ohio.jpg/CC BY 2.0 Generic; p. 4 Elena Masiutkina/Shutterstock.com; p. 10 Look and Learn/ Bridgeman Images; p. 13 Aunt Spray/ Shutterstock.com; p. 15 Norman Einstein/Wikimedia Commons/File:Fertile Crescent map.png/CC BY SA 3.0; p. 20 Hulton Archive/Getty Images; p. 21 Spencer Sutton/Science Source/Getty Images; p. 24 Bettmann/ Getty Images; p. 28 Tao Jiang/EyeEm/Getty Images; p. 31 Udaix/Shutterstock.com; p. 35 Sheila Terry/ Science Source; p. 36 Comstock/Stockbyte/Thinkstock.com; p. 40 Alexlky/Shutterstock.com; p. 48, 51 Science Source/Getty Images; p. 53 Karine Aigner/National Geographic Magazines/Getty Images; p. 57 Classic Image/Alamy Stock Photo; p. 60 Joseph Scherschel/The LIFE Images Collection/Getty Images; p. 64 Zeljko Radojko/Shutterstock.com; p. 68 V. Habbick Visions/Science Source; p. 70 Luke Sharrett/Bloomberg/ Getty Images; p. 73 Jim McKnight/AP Images; p. 75 Scimat/Science Source; p. 84 By 360b/Shutterstock.com; p. 89 Universal Images Group North America LLC/Alamy Stock Photo; p. 90 F Lariviere/Shutterstock.com; p. 95 David Grossman/Alamy Stock Photo; p. 98 Ira Bostic/Shutterstock.com; p. 101 Gideon Pisanty/ Wikimedia Commons/File:Lolium rigidum 1.JPG/CC BY 3.0; p. 104 Alison Hancock/Shutterstock.com.

Printed in the United States of America

Contents

Corn is a major player in genetically modified crops. About 88 percent of US-grown corn is genetically altered.

Introduction

Agriculture is one of the oldest human pursuits. The practice of breeding plants, animals, and other organisms for food and other human products stretches back ten thousand years. Today, agricultural land comprises 38 percent of Earth's total land mass. Historians, biologists, and other scientists all contributed to our understanding of the ways in which early humans developed practices that still influence agriculture today.

For thousands of years, humans have selectively bred plants and animals for offspring with more desirable characteristics. Farmers chose plants and animals with the best traits, like faster growth or hardiness. They bred these organisms over weaker versions. Organisms with favorable traits resulted in stronger and more productive crops and livestock.

The enormous differences in dog and horse breeds, for example, are the result of thousands of years of selective breeding. The modern-day golden retriever can fetch, or retrieve, birds for hunters. During the mid-1800s, hunters found that existing dogs were not adept at retrieving birds from water and land. Hunters needed a dog breed that could swim and fetch birds in marshes and wetlands. Scottish hunters bred water spaniels with existing retriever breeds to form a new type of dog, the golden retriever. Golden retrievers are excellent swimmers and can retrieve downed birds.

Over time, our understanding of how traits, like swimming, transfer from parents to offspring grew. As scientists discovered the principles of heredity, they developed ways to change organisms' traits. Using this knowledge, scientists created genetically modified organisms (GMOs).

To create genetically modified plants and animals, scientists combine the genetic material of different, often unrelated, species. The genetic information can come from other animals, plants, bacteria, and even humans. With technology to analyze genetic information, scientists identify traits that will be beneficial to other organisms. Then, they use different genetic modification, or change, techniques to combine the genetic information of two different species.

One of the most common uses of such technology is the production of genetically modified crops (GMCs), such as corn, rice, apples, squash, and potatoes. Using traits from bacteria and other organisms, scientists have designed GMCs that can produce higher yields. They have also created genetically modified crops that can survive challenging environments.

Genetically modified crops are widely planted in the United States and globally. Between 1996 and 2011, farmers from twenty-nine countries planted 1.25 billion hectares (3,088,817,268 acres) of genetically modified crops, an area of land greater than the United States or China. The United States dedicates the largest area of land, around 70 million hectares (172,973,767 acres), to GMC production.

Today, there are over forty types of genetically modified plants. The most common are corn, soybeans, cotton, and canola. Other GMCs include tobacco, sugar beets, and potatoes. Many ingredients found in packaged food come from GMCs.

Genetically modified crops are also grown for livestock feed. These crops are more resistant to insects, pests, and weeds. Because of their resistance, farmers use less pesticides and herbicides. They are able to survive drought and extreme temperatures. Changing the genetic makeup of these crops also creates tougher, higher-producing plants. Genetically modified corn, cotton, and soy, for instance, produce 20 to 30 percent higher yields than nonmodified crops. Higher yields lower food production costs and provide more food. More food at lower costs is critical for many who face hunger across the globe.

Despite their popularity, GMCs are controversial. Supporters see genetically modified crops as the solution to hunger as the global population grows. Critics argue that genetically modified crops and animals may have harmful effects on human health and the environment. Critics also believe that there may be dangerous long-term effects. Genetic modification may, they point out, cause allergies or alter the nutritional value. Supporters, however, maintain that genetically modified crops offer enhanced nutritional value in addition to a longer shelf life, which is beneficial for consumers.

There is also concern that genetically modified crops may alter the environment. For example, "superweeds," or herbicide-resistant plants, have developed. The problem with superweeds is that they require greater herbicide use, which can be harmful to humans, other animals, and the environment.

In fact, there is a consensus in the scientific community that currently available genetically modified foods do not pose a threat to human health. However, every new genetically modified crop must be tested independently. Critics also argue that genetically modified crops must be evaluated on their environmental, social, and economic impacts. These impacts constitute three foundational components of sustainable agriculture.

One company, in particular, is often at the heart of the controversy surrounding genetically modified crops. Monsanto Corporation produces about 90 percent of the world's genetically modified crops using a combination of selective breeding techniques employed by farmers for centuries and cutting-edge genetic technology. In addition to lawsuits against farmers, questions about unfair business practices and herbicide production have all lent themselves to the company's disputed reputation.

World governments vary in their legal treatment of genetically modified crops. Most European countries require that any product containing genetically modified crops and intended for human consumption be labeled. Currently, the United States has no labeling laws. Even so, the controversy is large enough that many companies and grocery stores do not use genetically modified crops in their products. General Mills removed GMCs from Cheerios, for instance, in 2014. Other companies, like the ice cream producer Ben & Jerry's, have publicly denounced the use of genetically modified crops in their products.

Are genetically modified crops the answer to ridding the world of hunger and starvation? Are they key to growing crops in challenging environments? Will genetically modified crops ensure everyone has access to affordable and nutritious food? Or, do they pose too great a risk to the environment and sustainable agriculture practices? Are scientists disrupting natural processes for the sake of lowering crop production costs?

In this book, discover how humans have altered the traits of plants and animals for centuries. Learn about the early scientific explorations that helped advance our understanding of genetics, and explore the genetic engineering methods utilized to create GMCs. Read about the intense debate over genetically modified crops, and decide what you think!

Scientists believe that the first agricultural revolution happened around 10,000 BCE.

CHAPTER 1

The Agriculture Problem

In order to understand the importance of genetically modified crops, one must journey back to the end of the Pleistocene epoch, or the Ice Age, about 11,700 years ago. Temperatures across the globe rose, melting glaciers and changing ecosystems. The change in temperature resulted in more temperate climates across the globe. The Pleistocene epoch began approximately 2.6 million years ago. During this most recent Ice Age, glaciers covered much of the planet, including parts of Europe, Asia, and North and South America. Remnants of the glaciers can still be seen today in Greenland and in northern parts of the North American continent. Although many Ice Age animals, like woolly mammoths and saber-tooth tigers, went extinct around thirteen thousand years ago, mammals that we are familiar with today, including apes, cattle, bears, and deer, were present during the Ice Age. Vegetation, while scarce, included conifers and grasses. During the Pleistocene epoch, modern humans, or *Homo sapiens*, originated. Near the end of the epoch, humans could be found across the globe. Humans and other hominid species like Neanderthal existed as hunter-gatherers, moving from place to place in search of food. As the planet warmed, large swaths of ice melted, leaving nutrient-rich soil and warmer temperatures, allowing humans to adapt their hunting-gathering practices to include the domestication of plants and animals. Domestication is the process by which wild animals or plants become adapted to humans and the environment that they provide. Organisms are selected based on desirable traits, and this leads to changes over time. Domestication occurs in areas that have a lot of biodiversity.

A CHANGING WORLD

The move from hunter-gatherer lifestyles to the domestication of plants and animals, or agriculture, is called the first agricultural revolution. Scientists once believed that humans began to manage various plants and animals as a response to increases in their population. However, modern evidence suggests that the development of agriculture was more of a way to supplement the lean availability of food during harsher months. Foraging for food has been the occupation of humans for 90 percent of our existence. Traveling to find edible vegetation and to hunt is a practice that has persisted in some cultures into the present day. Around eighty thousand years ago, bands of humans began concentrating on hunting and gathering more specific types of plants and animals. They created specialized tools like fishing hooks and harpoons. As early as twelve thousand years ago, humans were practicing forest farming. Humans

Evidence suggests that early bands of hunter-gatherer humans hunted woolly mammoths before settling into more permanent communities.

identified edible plants and aided in their protection and cultivation. This may be one the earliest examples of human influence on the growth of other organisms. In fact, many of the common vegetables known today are so different from any wild species that scientists do not know from which specific ancestors they evolved. Early humans probably selected perennials for forest farming and domestication. Perennials are plants that live for several years and produce seeds or flowers over multiple seasons. By placing parts of a plant, like the roots or leaves, in the ground, early farmers could grow new plants. This is called vegetative reproduction. Plants that are replanted each year are called annuals. Early humans saved the seeds of annuals and planted them each new season. Over time, these ancient farmers learned to plant seeds or other parts from only the very best producing crops, leading to better, stronger versions over time.

It is important to note that domestication and wild growth differ in several ways. First, competition between and among species is high in the wild. Plants must compete for light, water, and nutrients. Competition is not as necessary for domesticated plants as humans tend to them and provide these highly sought-after resources. Second, production size is often the main concern for domesticated plants, whereas reproduction is the priority in naturally occurring environments. Lastly, over time, humans exerted influence on modern fruit and vegetable varieties not only by choosing what they planted, but also by choosing what they consumed. Demand for vegetables resulted in different crop varieties that would not have occurred without human intervention. For example, cabbage, kohlrabi, broccoli, cauliflower, and brussels sprouts are all descendants of one type of wild cabbage.

Archaeological evidence indicates that domestication of plants and animals resembling modern agriculture began in southwest Asia, in an area called the Fertile Crescent. Surrounded by three rivers, the Nile, the Euphrates, and the Tigris, the soil in this half-moon-shaped region is particularly rich in nutrients. The area serves as a land bridge between Africa, Europe, and Asia.

Human-like ancestors occupied the Fertile Crescent before modern *Homo sapiens* evolved. Near the end of the Pleistocene, around 12,000 BCE, a group of *Homo sapiens*, the Natufian culture, inhabited the area. They subsisted primarily by hunting gazelles and gathering wild grains.

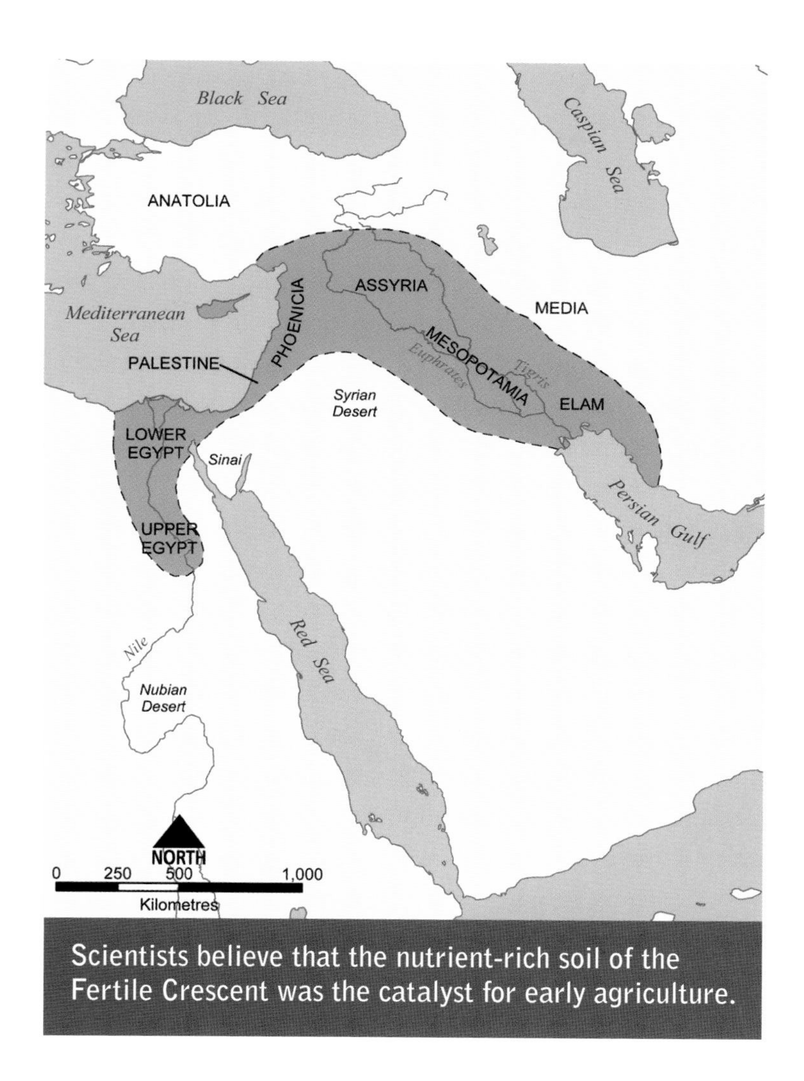

Scientists believe that the nutrient-rich soil of the Fertile Crescent was the catalyst for early agriculture.

They were unusual for human groups at the time as they were more sedentary than most hunter-gatherer groups of the Ice Age. Anthropologists and archaeologists know these details about the Natufians due to the discovery of grain-harvesting tools. Around the same time, other human groups in the eastern Fertile Crescent, near modern-day

Iran, domesticated sheep and goats as a supplement to hunted gazelles. Evidence suggests that the domestication of plants soon followed. Some scientists argue that a drought, or prolonged period without rainfall, forced early humans to cultivate the plants they were no longer able to gather in the wild. Not all members of the scientific community accept this theory. Simultaneously, humans in east Asia domesticated grain and rice. Around 8000 BCE, inhabitants of the American continents grew squash. The domestication of plants in these various places all happened independently of one another.

Available land for early agriculture was small because of the limited tools humans had. Often, they had only sticks and wooden hoes, and heavily wooded or brushy areas were difficult to manage. Sometimes, early humans also used fire to burn down wooded areas. As early agricultural techniques developed, humans learned that growing larger quantities of plants could, over time, remove nutrients from the soil. Early farmers began employing fallow farming in which part of the ground was left to rest during a growing season.

Abu Hureyra, where modern Syria is today, is the largest site where evidence of early agriculture has been found. Scientists found evidence that the people in that region raised lentils and rye and eventually expanded to more crops like legumes and wheat. In Central, North, and South America, humans also cultivated plants.

While there are multiple sites in which early agricultural practices developed, domestication practices spread and gave rise to large settlements of people. The Sumerians in ancient Mesopotamia lived in large cities and developed agricultural technology, such as plows. Situated between the

Euphrates and Tigris Rivers, the yearly flooding allowed crops to flourish in the dry environment. Surplus due to the flooding allowed the Sumerians to settle in one place. Their most important crops were wheat and barley, but Sumerian farmers also cultivated grapes, melons, and eggplant. Their civilization is responsible for many technological inventions still used today. They invented bronze axes and sickles. They are also believed to be responsible for inventing the wheel. Around 3000 BCE, there is written Sumerian record of the invention of the plow. The plow can be human, animal, or in modern times, machine powered. Plows turn the soil and dig indentions for seeds. The Sumerians also used irrigation to move the rivers' floodwaters into their fields.

Between 10,000 and 4000 BCE, the ancient Egyptian empire became one of the first to practice large-scale agriculture, growing wheat, barley, flax, and papyrus in the fertile soils alongside the Nile River. Simultaneously, the ancient Chinese documented their agricultural practices. Scientists believe that Asian rice was domesticated from wild rice approximately 12,500 years ago and subsequently spread throughout southeast Asia and the West. In addition to writing about their agriculture practices, the Chinese utilized animals such as oxen to aid them in tilling and tending their crops.

The Roman Empire is another civilization that left many written records on agriculture. In addition to advancing warfare and conquering technology and tactics, the Romans also prized agriculture. Land ownership was incredibly important to securing one's social standing in ancient Rome. Borrowing agricultural practices that are still in use today, like crop rotation, from the Greeks, Roman farms expanded from small family-held parcels to sprawling estates owned by the aristocracy. Farms were typically organized in one of three ways: small family farms, tenant farming, or slave labor. Tenant farming is when a landowner provides land and tenants, or sharecroppers, perform the labor of planting, tending, and harvesting.

During the Middle Ages, from 100 BCE to 1200 CE, agriculture technology continued to advance. In the ancient Islamic world, Muslim traders sold crops like sugar cane, rice, and cotton along trade routes. In medieval Europe, land ownership and social classes resulted in the serfdom system. The Middle Ages were a tumultuous time of war, violence, and changing power. Individuals called serfs worked the land for aristocrats, like knights, barons, or lords, and, in return, they were provided protection by the landowner. Serfs could cultivate smaller portions of the land within the manor for their own food. The prevalence of serfdom declined around the fifteenth century. Some scientists attribute this decline to the black death, or the bubonic plague, which resulted in the deaths of an estimated seventy to two hundred million people in Europe and Asia.

Prior to the eighteenth century, agriculture on the British Isles remained relatively small scale. In 1750, the population quickly grew to 5.7 million, and agricultural yields could not keep up with demand. During what was called the British agricultural revolution, agricultural practices and technology advanced dramatically between 1750 and the mid-nineteenth century. Major developments included crop rotation, enclosure, increases in farm size, and selective breeding of livestock animals.

Crop rotation is the practice of growing different crops in the same area during different seasons to restore nutrients to the soil. Different plants require different levels of common soil nutrients, including nitrogen, potassium, and phosphorous; thus, when planted, they absorb different quantities of each. Crop rotation ensures healthier soils, and it allowed British farmers to produce more on the same plots of land.

Enclosure refers to the practice of removing public access to privately owned land by fences or other barriers. Remnants of Middle Age serfdom practices lingered into the eighteenth and nineteenth centuries. Often, those who farmed for survival did so on land owned by the church or higher members of British society. Allowing subsistence farmers access to privately owned land was referred to as the open field system. Enclosure began around the time of the black death and continued into the eighteenth century. Enclosed land required fewer people to tend it and left many poorer British without access to land to grow food. With these cultural changes came migratory changes. The poor moved to cities to seek work in the new factories that emerged because of the Industrial Revolution. Others sought their fortunes in the New World, leaving on massive ships with the hopes of land ownership and wealth in British-held colonies, including much of North America. During the time of enclosure, the average size of farms increased due to advances in technology that allowed more land to be farmed, even if the land was not naturally fertile. Lastly, in the mid-eighteenth century, British farmers began selectively breeding livestock, mainly sheep.

By choosing male and female animals with the best traits, British farmers bred sheep specifically suited for wool production or meat. All these advances allowed agricultural production in Britain to soar, even surpassing the population boom.

The Father of Genetics

Although by the nineteenth century, humans had been selectively breeding different plant varieties for thousands of years, scientific plant breeding required an understanding of heredity and genetics. In 1865, an Austrian monk named Gregor Mendel conducted a series of experiments that would change both agriculture and science forever. A student of math and physics, Mendel was asked by the abbot of his monastery to study selective breeding. Although sheep were the originally intended subjects, due to the desire to increase the monastery's wool profits, Mendel ultimately chose plants, specifically peas, to study. Pea plants were the ideal experimental subjects because their traits are easily observed.

Gregor Mendel is considered the father of genetics.

They also grew quickly and produced new generations rapidly for study.

To fully understand Mendel's pea experiments, it is important to first understand the processes of seed germination and pollination. Seed germination refers to the process by which a seedling sprouts and eventually becomes an adult plant from a seed. Pollination is the way plants reproduce. Pollination occurs when insects and

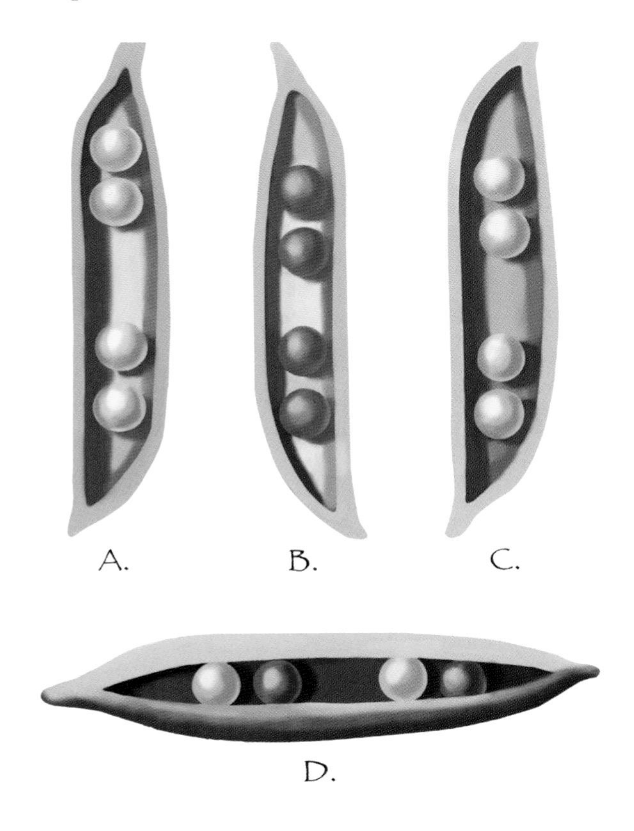

Mendel's pea plants had two seed color varieties: green and yellow.

other environmental factors, like wind, transport pollen, or the male reproductive cells of plants, from one plant to another. The pollen fertilizes egg cells of another plant, and those eventually form seeds. Plants can also self-fertilize, or produce seeds from their own reproductive cells. Self-pollination and cross-pollination will be examined in more depth later on. Humans can manipulate pollination and germination by controlling which seeds are planted and by transferring pollen from one desired plant to another.

Mendel chose to study pea plants because they grow quickly and have traits that are easy to observe, such as seed color, texture, and height. Pea plants can produce yellow or green seeds. Pea seeds can also be wrinkled or smooth. Both of these traits are passed down from one generation to the next. Through cross-pollination, Mendel bred pea plants with different traits, such as height, seed color, or texture. Often, the offspring would only have the traits of one of the parent plants. In his first round of experiments, for example, Mendel cross-pollinated plants with yellow seeds with green-seeded plants. All of the descendent plants had yellow seeds. The green seed trait disappeared. He then bred two of the yellow offspring. About 75 percent of the third generation had yellow seeds and about 25 percent had green seeds.

Mendel continued to breed successive generations of pea plants. In each generation, the ratio of yellow seeds to green seeds remained fairly consistent. About three-quarters of the offspring had the yellow seed trait and a smaller percentage, about one-quarter, had green seeds. The results were similar with seed texture: about 75 percent of the offspring had smooth seeds while only 25 percent had wrinkled seeds. Even when Mendel cross-pollinated plants with different combinations of traits, these ratios remained constant. Mendel used these results to develop several important rules for how organisms acquire traits from their parents. This field of scientific study is called heredity.

Mendel published his findings in two lectures in 1865 to the Natural Science Society and published a subsequent paper called "Experiments on Plant Hybrids." At the time, both Mendel's lectures and paper received little attention, and, in fact, it wasn't until the twentieth century that scientists revisited his ideas. Mendel's discoveries of trait inheritance and the concepts of trait dominance would forever alter the study of genetics and pave the road for genetically modified crops.

Modern Agriculture

Modern agriculture is all about production. Through the use of machinery like tractors, combines, and harvesters, humans have increased crop yields. Monoculture, or growing one crop across large areas, is common practice. Despite all the advances that have characterized the history of agriculture and created modern agriculture, hunger is still a challenge faced by many in the world. Some 735 million people do not have enough food to lead a healthy and active life.

The use of pesticides to increase yields in the twentieth century has also threatened the environment. Humans have practiced various forms of pest and weed prevention since the early domestication of the Fertile Crescent. Pesticides are used to kill organisms that threaten the growth of crops. The most common pests are weeds, insects, and fungi. Pesticides

Natural events, like the dust bowl of the 1930s, can lower crop yields, causing illness and starvation.

are generally chemical agents. The Sumerians first used sulfur to ward off harmful insects. By the fifteenth century, deadly elements like arsenic and mercury were applied to crops.

In the early twentieth century, inorganic substances, or substances that do not naturally occur, grew popular. By-products of the coal industry, these chemicals allowed for massive gains in production, lowering costs for consumers. The most popular of these was

Dichlorodiphenyltrichloroethane (DDT). Used against insects, DDT became available for sale in 1945, and it was promoted by both private corporations and the government. In 1962, Rachel Carson published her book *Silent Spring*, which argued against the use of these inorganic chemical pesticides because of the harm they did to beneficial insects and habitats. The book also argued that DDT and other pesticides caused cancer. The publicity storm surrounding Carson's book and subsequent research led to the banning of DDT in the 1970s.

Although scientists continued to research methods of pest control that were safer, insects and weeds still continued to have negative effects on crops. Weather and environmental events are also modern threats to crop production, even with such advances in agricultural technology. Using Mendel's and other scientists' discoveries, researchers in the latter half of the twentieth century would explore solutions to the challenges that have plagued farmers since the dawn of agriculture.

Disasters of History

Despite the many incredible agricultural innovations of ancient civilizations, humans have always had to contend with natural events out of their control. Droughts, floods, storms, and pests have all negatively impacted crop production throughout history. For instance, although ancient Egyptians successfully grew a diverse number of crops in the rich flood lands of the Nile, anthropologists and scientists believe that a famine may have caused the fall of the great empire.

Experts believe that the terrible hunger, documented by Egyptian scholars around 2180 BCE, was caused by a reduction in the flooding of the Nile River. Without the nutrient-rich soil, farmers were unable to produce enough food for the population. This may have weakened the civilization, allowing it to be vulnerable to outside invaders.

Such calamities also happened in the nineteenth century and even in recent history. When, for instance, a fungus destroyed the potato crop in Ireland, it decimated the Irish population. This disaster is called the Great Famine because, at the time, two-fifths of the Irish population depended on the potato as the primary staple of their diet.

Many Irish died, and others fled to the United States during a period from 1845 to 1849, causing Ireland's population to drop by approximately 25 percent. In the 1930s, the plains of Canada and America were struck by severe dust storms caused by drought and, in part, by farming practices that eroded the soil.

Tens of thousands of poor farmers and their families fled what became known as the great dust bowl. They were unable to grow crops or pay their bills. Children wore dust masks, and farmers watched the destruction of their crops. While the dust and drought primarily impacted the plains in the South, the impact was also felt in the North. The agricultural devastation actually created even more challenges for the North during the Great Depression. Environmental disasters have always challenged farmers and their families and, in some cases, have been deadly.

A mule is bred from a female horse and a male donkey.

CHAPTER 2

The Science of Genetically Modified Crops

Recall that selective breeding is the process in which humans select animals or plants to breed based on naturally occurring differences in their traits. Selective breeding was a foundational practice for early agriculture and, subsequently, early civilizations. By selecting crops or livestock with the best traits, early humans domesticated crops, raised animals for their meat and other products, and settled in communities larger than ever before. Variations between organisms of the same species are the driving force behind natural selection. The mechanism that drives evolution, natural selection is the process in which organisms with traits best suited to their environments survive and pass on their genetic information to their offspring. Plants, animals, and other organisms have genetic differences due to the different combinations of traits passed from one generation to the next. Think of humans. A child receives half of his or her genetic information from one parent and half from the other. This results in a new individual with a completely different set of genes. All humans, except identical twins, have different genetic codes. Variations in genetic codes can also be caused by changes in an organism's DNA (deoxyribose nucleic acid), called mutations. These changes can occur because of mistakes in an organism's genetic blueprint or because of influences in the environment that cause changes in DNA. Mutations are discussed more thoroughly later in this chapter. Early humans capitalized on these naturally occurring processes to domesticate other species.

HYBRIDIZATION

Millennia after civilizations in the Fertile Crescent began selectively breeding plants and animals, humans took another step and began creating entirely new species. By crossbreeding two different species, which is known as hybridization, humans could create organisms that were not naturally occurring. The results of hybridization can be seen today. For example, mules were an early hybrid. When a male donkey is bred with a female horse, the offspring is called a mule. Mules were depicted in early Sumerian art. There is evidence that two types of rice may have been the first crossbred plants. Around 2000 BCE, a species of rice from what is now modern-day Japan may have been crossed with a species of rice from the sub-Asian continent, forming an entirely new species.

Often, different species were bred to obtain a new and useful trait from the combination of the two parents' characteristics. This is called intraspecific hybridization. Another type, interspecific, occurs when two members of the same species are interbred to improve the species. Corn, for instance, was interbred over the course of millennia so the modern-day crop looks nothing like what the original ancestor plant did. Dogs, cats, and roses are all other examples of hybridization—each breed the result of a combination of two different species.

Hybridization results in new trait combinations. This is true for both physical and nonphysical traits. Physically, mules, for instance, have small hooves, large ears, and short manes like donkeys. Their body shape, though, is horse-like. From the male donkey, mules inherit sure-footedness, or the ability to walk on rough terrain, and endurance. From the female horse, mules inherit speed and agility. Because of the

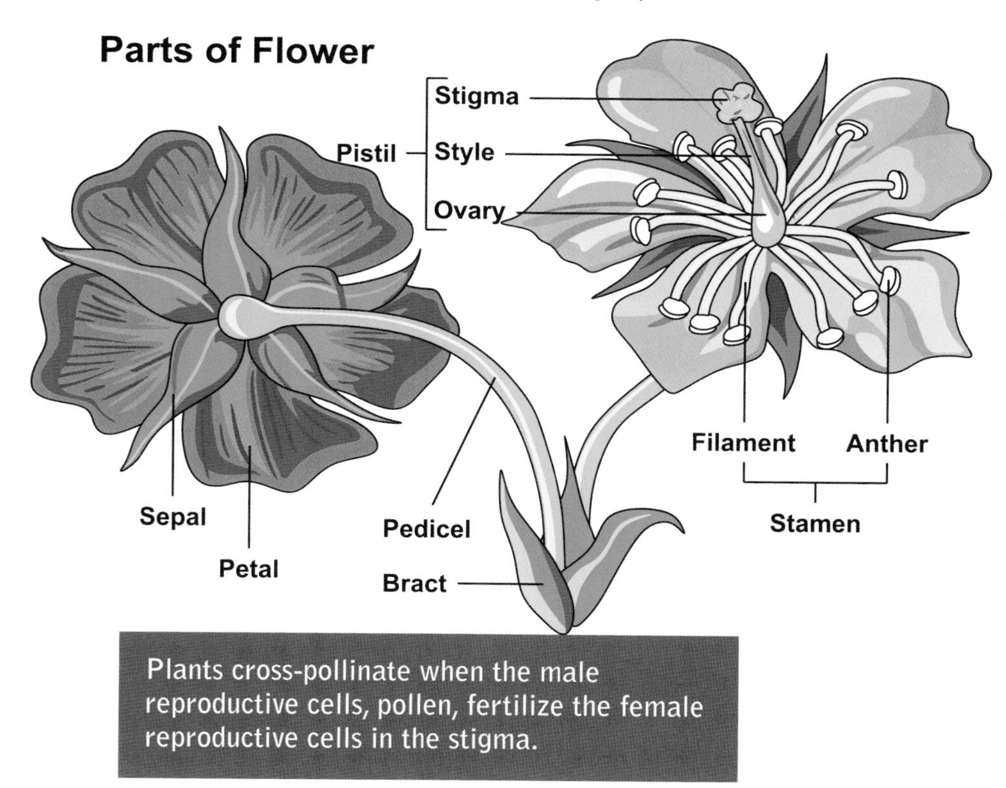

Plants cross-pollinate when the male reproductive cells, pollen, fertilize the female reproductive cells in the stigma.

combination of their parents' traits, mules are used to pack large loads over rugged ground.

In plants, hybridization can be accomplished with several methods. The first, cross-pollination, was used by Gregor Mendel to breed pea plants with specific traits. Recall that pollen is the male reproductive cell of plants. Pollen is transferred from the male part of the plant to fertilize the egg, or ovule, in the female plant part, or stigma. This process produces offspring, or seeds, that will grow into new, independent plants. The fusion of reproductive cells from two separate organisms forms offspring genetically different from either parent. Cross-pollination occurs when a vector transfers the pollen of one plant to another. A vector can be naturally occurring, such as an insect or wind carrying pollen. Animals that assist in this method of plant reproduction are called pollinators. Pollinators include butterflies, ants, birds, moths, and bats. Cross-pollination can also occur artificially as performed by Mendel in his pea plant experiments. Mendel specifically selected plants based on their characteristics and then cross-pollinated them.

Today, plant scientists and farmers cross-pollinate new generations of plants with desirable combinations from two parents with the most successful qualities. Breeders can then test seeds under adverse conditions to ensure that they have the best possible traits.

Another mode of plant hybridization is grafting. Grafting combines two separate plants into a single organism. As opposed to cross-pollination and the production of a hybrid seed, grafting is when the root structure of one plant, called a rootstock, is attached to the top portion of another plant, called the scion. Using rootstocks allows farmers to grow plants more quickly and in difficult conditions. Rootstocks also contribute traits that cause plants to be more productive, resistant to environmental hardship, and, in some cases, resistant to diseases. The top portion, the scion, is a bud, or shoot. A scion comes from a young plant that has desirable characteristics like excellent color, flavor, or resistance to diseases. Rootstocks and scions do not have to be from the same species but must be closely related, like lemon and orange trees. In fact, most of the fruit trees in America are grafted onto rootstocks to increase their production and improve certain traits. Early farmers first domesticated fruit plants like figs, grapes, and olives by simply growing new plants from offshoots or cuttings. Grafting was not discovered for several thousands of years. Around 1000 BCE, farmers began using the techniques to grow fruit, such as apple, pear, and plum trees. Such technology allowed fruit tree cultivation to spread from central Asia to Europe. Both grafting and cross-pollination unite the genetic information of two different species and make the resulting organism better.

Mutagenesis

Another way in which new traits arise in plants is through genetic mutations. Mutations are changes in an organism's DNA that lead to new traits. Mutation can naturally occur or can be caused by environmental influencers, known as mutagens. Mutations can happen in all living things and may be passed on to subsequent generations. Blue eyes, for instance, are the result of mutated human genes that were passed down for thousands of years.

While mutations are naturally occurring, modern plant breeders can induce mutations through the introduction of chemicals that alter plant DNA. They can then select plants with desirable, mutated genes.

For over three thousand years, farmers have used hybridization techniques like breeding different species, cross-pollination, and grafting to improve the quality of their crops and animals. Although they did not understand the scientific basis behind the process of hybridization, through trial and error, they successfully bred hybrids like peppermint (a cross between spearmint and water mint) and the beefalo (cow and buffalo). Even though thousands of years of hybridization took place without understanding the underlying genetic processes, discoveries made by scientists like Gregor Mendel would pave the way to modern genetic engineering.

The STRUCTURE of GENES

Through his pea plant experiments, Gregor Mendel hypothesized the rules by which organisms pass on their genetic information. These rules are called Mendel's laws of heredity. First, his experiments demonstrated that organisms inherit different characteristics, like seed color or texture, independently of one another. An individual trait is inherited as a pair—one piece of information from the father and one from the mother. This rule, that offspring inherit two copies of each trait's instructions, is called the law of segregation. The second rule is the law of dominance. Although an organism inherits traits as pairs from its parents, it will express the dominant trait,

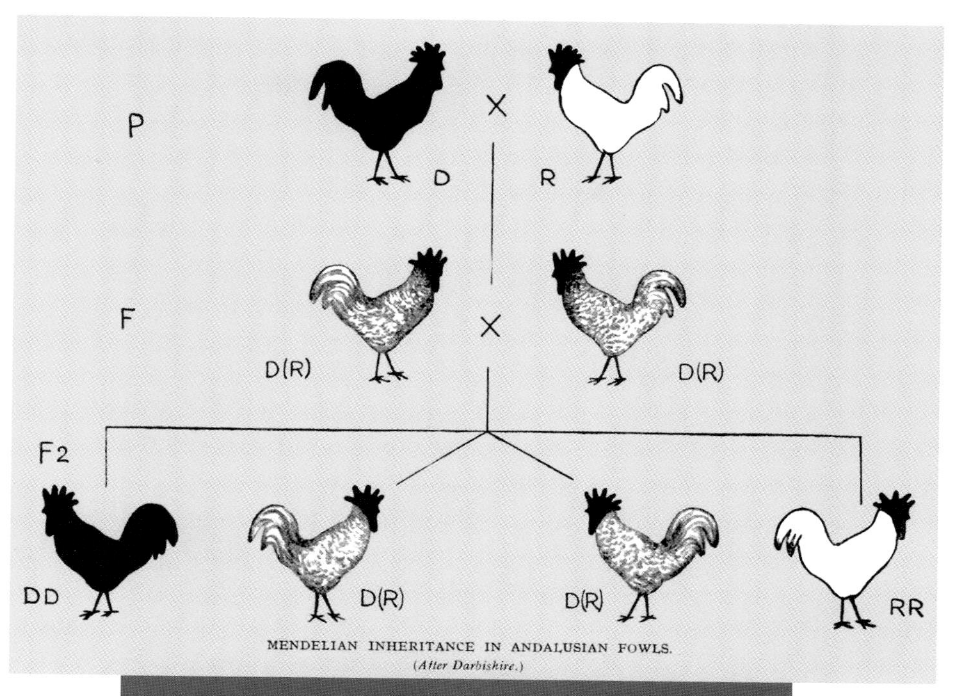

Beyond basic inheritance, organisms can also pass a mixture of dominant traits, such as feather color, to their offspring.

if inherited. For instance, Mendel observed that in each subsequent generation of pea plants, those with yellow seeds outnumbered plants with green seeds by a three-to-one ratio. This pattern persisted through multiple generations. The yellow seed trait is dominant over the green seed trait. Because the pea plants inherited seed color as a pair, as long as the genetic instructions for yellow seeds came from one parent, the trait expressed was yellow. Even if a trait is not physically present, the information for a masked trait, like green seed color, will still be passed down from parent

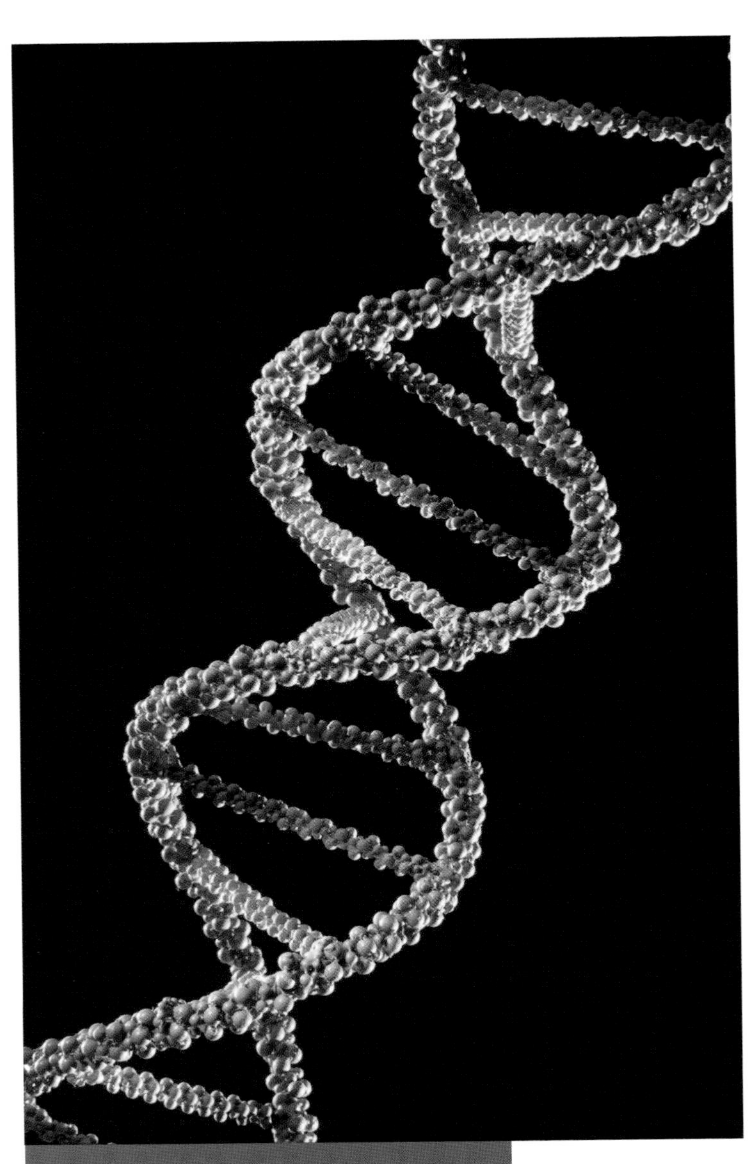

Deoxyribonucleic acid (DNA) contains the instructions for protein production in living things.

to child. Mendel coined the term "recessive" to describe masked traits like green seed color. A pea plant must receive copies of the green seed trait from both parent plants in order to express that trait. Lastly, Mendel's experiments demonstrated that traits are inherited independently. The law of independent assortment states that the inheritance of one trait will not affect the inheritance of other characteristics. For instance, the trait of eye color does not influence the inheritance of hair color. Gregor Mendel's work postulated that all genetic information was encoded in some unknown substance, which Mendel called "elements." When Mendel published his findings explaining the process through which living things inherit traits from their parents in a uniform and predictable manner, they largely went unnoticed for approximately three decades.

Mendel was, in fact, absolutely correct in his theory that genetic information is carried and transferred between parent and offspring in a single, uniform chemical. DNA is the blueprint of all living things. In humans, plants, and animals, DNA is found in the nucleus of every cell. The genetic information of DNA is stored in four chemical units called nitrogen bases: adenine (A), thymine (T), guanine (G), and cytosine (C). The sequence of these bases determines the genetic information used to determine all of an organism's characteristics. Much like the order of the letters in this sentence forms its meaning, the order of nitrogenous bases determines the unique composition of each and every living thing. The four DNA bases pair with one another—adenine and thymine; guanine and cytosine—to form base pairs.

These pairs attach to a sugar molecule and a phosphate molecule to form a nucleotide. Nucleotides form a double helix that resembles a twisted ladder. The base pairs resemble a ladder's rungs, and the sugar and phosphate molecules form the ladder's sides. Human DNA contains about three billion bases! Sequences of base pairs form genes, the basic unit of heredity. The instructions for all traits are contained in genes. Every organism inherits two versions of every gene, one from each parent.

How do genes become traits like seed color or height? The discovery of this process is credited to Dr. Marshall Nirenberg. By the late 1950s, scientific experiments identified DNA as the basic unit of heredity. Scientists did not know, however, how the chemical information stored in the nitrogen base pairs translated into discernable characteristics. They knew that the information stored in DNA was first transcribed, or transferred, to a chemical known as RNA. Somewhat similar to DNA, ribonucleic acid is a single-stranded unit that uses uracil (U) instead of thymine. Nirenberg and colleagues discovered that the information stored in RNA is then translated into amino acids, which are the basic chemical building blocks of all living things. Amino acids form long chains called proteins. Dr. Nirenberg discovered that through this process of DNA transcription and translation, the genetic code contained in DNA is translated into proteins. Proteins perform a variety of functions in the body, including composing biological materials and running biological processes.

GENETIC ENGINEERING: HISTORY and PROCESS

Understanding the basic structure of DNA, the process through which it replicates, and the way that genetic information is translated into proteins is key to understanding the need for genetically modified crops. A plant, for instance, that is able to thrive in extremely dry environments can do so because its DNA contains genes that code for traits that allow it to tolerate droughts. If a plant is unable to tolerate drought, however, its DNA does not contain the correct series of nitrogen bases for low water environments. What if scientists could change those bases? Or, what if they could insert the correct bases for those traits? This is the process of genetic engineering.

In order to change the DNA of one organism to contain traits of another, scientists must first understand the genes contained in an organism's entire genetic code, or genome. Mapping and sequencing genomes is a key building block to being able to alter and change an organism's genes.

In 1909, Dr. Thomas Morgan began his famous work with the common fruit fly, named *Drosophila melanogaster*, at Columbia University in New York City. Dr. Morgan and his students studied the fast-breeding fruit fly in his laboratory, called the "fly room." Morgan wished to understand the mechanisms through which Mendel's pea plants inherited their various traits. By studying thousands of fruit flies with magnifying glasses and microscopes, Morgan and his students discovered much about the

Fruit flies were key to Thomas Morgan's discoveries about chromosomal structure.

physical arrangement of genomes. Through his experiments, Morgan hypothesized that genes are arranged in a linear structure on chromosomes, like beads on a string.

Dr. Alfred Sturtevant, a student of Dr. Morgan's, created the first genetic map in 1913. As a student, Sturtevant worked with Dr. Morgan in the fly room. Sturtevant studied three traits of fruit flies: their eye, wing, and body characteristics. Through the study of multiple generations, he determined how often these traits were inherited together. Using that data, Sturtevant created a map of the fruit flies' genes. Gene mapping was a precursor to later efforts to sequence an entire genome, or determine the sequence of all nitrogen bases in an organism's DNA, and understand the genes contained in an organism's genome. Before the invention of genome sequencing, scientists discovered another key process to future genetic engineering—cutting and inserting genetic information into a string of DNA.

Recombinant DNA

In 1972, Paul Berg set about carrying out a series of experiments on the DNA of two different viruses. Using naturally occurring chemicals that act as scissors and glue, Berg joined DNA sequences from the two different viruses. This results in a "chimera," a term from ancient Greek mythology because a chimera is a mythical animal formed from parts of various organisms. Berg's experiments were the first to use recombinant DNA, or the process of combining the DNA of two different organisms. Using a biological molecule called an enzyme, Berg cut out small sections of DNA from one of the viruses and inserted that segment into the other virus, producing a new organism with a combination of genes from both viruses.

At the same time that Berg was conducting his experiments, two of his contemporaries, Herbert Boyer and Stanley Cohen, were studying similar genetic processes in California. While Berg was able to form a chimera from combined viral genetic information, the next important step would be ensuring that that genetic information could be copied and passed on. Boyer and Cohen developed a way to ensure that DNA inserted into another organism's DNA would continue to replicate like it does naturally. Using a similar cut-and-paste method, Boyer and Cohen inserted foreign DNA into *E. coli* bacteria. The DNA coded for resistance to an antibiotic, tetracycline. They observed the bacteria's reproduction. Boyer and Cohen found that later generations of the *E. coli* bacteria displayed antibiotic resistant traits that are not naturally occurring in the *E. coli*. They used recombinant DNA technology not only to produce a new organism, but also to propagate a desired trait in subsequent offspring.

Finally, scientists needed to determine a fast method to copy DNA sequences in order to alter multiple segments. In 1983, Kary Mullis developed an inexpensive and cheap method of copying segments of DNA, called polymerase chain reaction (PCR). PCR is now one of the most popular techniques used in genetic science. Its purposes are broad and include cloning, genetic mutagenesis, genomic sequencing, and monitoring and analyzing genes. PCR is used in a variety of fields including medical research, crime analysis, and genetic research.

Organisms that have altered genomes, like GMCs, are called transgenic organisms. Transgenic organisms can have inserted genes from another organism or genes that have been disabled for the purposes of scientific research.

Rudolf Jaenisch, professor of biology at Massachusetts Institute of Technology in Cambridge, Massachusetts, and Beatrice Mintz, chair of the Fox Chase Cancer Center in Philadelphia, Pennsylvania, are considered pioneers in the field of genetic engineering. In 1974, Mintz and Jaenisch made a breakthrough discovery. They inserted virus DNA into embryonic mice. The mice were born with the DNA sequences in their genome and represented in the genes of their offspring. Both scientists received multiple awards and recognition for their contributions to the study of genetics and epigenetics. Epigenetics refers to modifying the way that an organism's genes are expressed as opposed to actually changing the DNA sequences.

The first transgenic plant was tobacco, produced in 1982. Scientists inserted DNA from bacteria that produce chemicals toxic to insects into the tobacco genome. These genes created insect resistance in the tobacco plants.

In order to create more complex transgenic organisms, like genetically modified crops, scientists first had to identify all the genes within a given organism's DNA and then understand which genes they coded.

GENOME SEQUENCING

The last decade of the millennium marked a race toward one of the greatest scientific achievements of the twentieth century. The goal was to sequence the entire human genome—approximately three billion base pairs and thirty thousand genes. In addition to the work of Mendel, Morgan, and Sturtevant, others made incredible discoveries that paved the way for the completion of the Human Genome Project in 2003.

In 1950, Erwin Chargaff determined that the nitrogen bases adenine and thymine, and cytosine and guanine were always present in DNA in equal amounts. From this, he deduced that adenine and thymine and cytosine and guanine always paired together. In 1953, James Watson and Francis Crick, based on work by Rosalind Franklin, discovered the double-stranded helix structure of DNA. Shortly afterward, in 1962, Dr. Nirenberg and his team made an impressive discovery. They cracked DNA's code by finding that DNA is ultimately translated into building blocks, or proteins.

In 1977, Frederick Sanger, eventual two-time winner of the Nobel Prize in Chemistry, developed a process to sequence smaller segments of DNA. Using his sequencing technique, Sanger and his team were the first to completely sequence a genome, that of a virus.

In 1990, the multinational Human Genome Project was launched. Scientists from all over the world contributed to the process of identifying the billions of bases encoded in human DNA. Meanwhile, scientists at The Institute for Genomic Research (TIGR) sequenced the entire genome of a bacteria, the *Haemophilus influenza*, in 1995.

The first animal genome sequenced was that of a nematode worm, the *Caenorhabditis elegans*. The genome of the fruit fly studied by Morgan and Sturtevant had its complete genome sequenced in 2000. The first plant genome to be sequenced was that of the *Arabidopsis thaliana*, a weed in the mustard family, in 2000. In 2002, the mouse was the first mammal to have its DNA completely sequenced. Scientists completed the human genome sequence the following year with approximately 99 percent accuracy.

Since the completion of the project, additional genomic sequencing feats have been accomplished. Chimpanzee DNA was fully sequenced in 2005. In 2008, the 1000 Genomes Project launched with the goal to sequence approximately 2,500 individuals' DNA.

Understanding the full genetic sequences of these organisms, and many others, agricultural genetic technology could move beyond simple selective breeding and hybridization. In 1982, scientists created the first genetically modified crop—tobacco. They inserted genes so that the tobacco plant would remain undamaged from the feeding of the larvae of a tobacco pest, the tobacco hornworm.

Frances Crick and James Watson

James Watson grew up in Chicago and received a bachelor of science in zoology from the University of Chicago in 1947. His childhood passion for ornithology, or the study of birds, morphed over time into a curiosity about genetics.

During his graduate studies at Indiana University, Watson was deeply inspired by geneticists on campus and began to study the effects that viruses had on DNA. Fascinated by this work on genetics, Watson set out to discover the structure of DNA.

Francis Crick's academic ambitions were interrupted by the outbreak of World War II in 1939. During the war, Crick designed weapons for the British military. Afterward, Crick returned to school and, while visiting Cambridge University, met a young scientist named James Watson. Even though Crick was twelve years older, they became good friends and colleagues.

Together, they began working in a laboratory at Cambridge University, researching the structure of DNA. Others laid the groundwork for their discovery, including Rosalind Franklin, a British chemist known for her pioneering work on DNA and X-ray diffraction.

Franklin earned her doctorate in physical chemistry from Cambridge University in 1945, despite the challenges of being a female scientist at the time. She worked on a research project that examined X-ray diffraction, a process used to develop structural pictures of molecules.

During her experiments, Franklin observed differences in X-ray results. At first, she believed that they displayed different molecules, but eventually she concluded that they showed different perspectives of the same molecule, DNA. Using the body of research, Watson and Crick built a double-helix model of DNA using paper and metal scraps. This incredible discovery led the pair to receive a joint Nobel Prize in 1963.

Francis Crick eventually became more focused on brain research. He also went on to become a professor at the Salk Institute for Biological Studies in California. He died on July 28, 2004. James Watson directed the Human Genome Project at the American National Institutes of Health from 1988 to 1992. Through this project, he encouraged corporations, the government, and scientists to work together.

Charles Darwin was twenty-two when he boarded the HMS *Beagle* for the Galapagos Islands.

CHAPTER 3

The Major Players

Although Charles Darwin is most often remembered as the father of the theory of evolution, it was botany, or the study of plants, that played an essential role in his life. Charles Robert Darwin was born on February 12, 1809, in Shrewsbury, Shropshire, England. He was the fifth of six children. His father, Robert, was a doctor. At age sixteen, Darwin worked as an apprentice for his father's medical practice. He went on to medical school in Edinburgh, Scotland. However, the medical field was not for the young scientists. While at school, Darwin studied taxidermy, or the practice of preserving animals by stuffing them. John Edmonston, a freed slave from the West Indies, mentored young Darwin and played a pivotal role in his understanding of the natural world. During his second year at university, Darwin joined a scientific debate group called the Plinian Society. The society often held lively debates in which they challenged long-held religious principles of creation. The Darwins were a religious family, and these debates were young Darwin's introduction to nonreligious scientific thinking. Around this time, Darwin also discovered the work of Jean-Baptiste Lamarck, a botanist and early proponent of evolution.

PLANTS, EVOLUTION, AND CHARLES DARWIN

In 1827, Darwin decided medicine was not his desired career path and dropped out of medical school. His father sent him to Christ's College to study theology. Again, Darwin was less interested in school and more interested in studying in the natural world. He spent great lengths of time outside, collecting insect specimens. Despite his skepticism about a career as a minister, Darwin passed his exams. However, he still expressed more interest in natural history than theology. He attended the lectures of Professor John Stevens Henslow, a botanist, geologist, and priest. The two grew close, and Henslow became Darwin's mentor. Henslow's career helped Darwin begin to see himself as a man of both faith and science. Henslow introduced Darwin to Professor Adam Sedgwick, a geologist at Cambridge, and he accompanied Professor Sedgwick on a geology trip to Wales. When Darwin returned home, he found letters from Henslow inviting him to be a naturalist on an expedition to South America. Sailing on the HMS *Beagle*, the scientific voyage would take two years. Darwin jumped at the opportunity.

He boarded the *Beagle* a few days before its departure on December 27, 1831. His quarters were the ship's chart room, at the stern, or back, of the ship. The room was cramped—about 9 by 11 feet (2.7 by 3.4 meters) and about 5 feet (1.5 m) in height. The ship set sail, and almost immediately Darwin became seasick.

The HMS *Beagle* was first launched in 1820, eleven years before Darwin's expedition.

Contrary to the original plan, the HMS *Beagle* would spend almost five years abroad. When the ship arrived in Brazil in 1832, Darwin hiked into the rain forest and returned with an enormous collection of insects and plants. Although he had spent much of his younger years observing and collecting specimens, he was, in many ways, a novice in naturalist studies.

He sent a collection of these specimens, along with careful notes and cataloging, back to Henslow. The *Beagle* then moved on to Patagonia, the southernmost region of South America, shared by Chile and Argentina. The area is known for its diverse climates due to the Andes Mountains.

The Argentinian side is grasslands and deserts while the Chilean side consists of temperate rain forests. Here, on Patagonia's coastline, Darwin collected many fossils. Despite the ship captain's misgivings about the volume of items he brought onboard, Darwin discovered several specimens that were not yet known to scientists back in England. In April 1834, they sailed around Cape Horn, and Darwin sent yet another shipment of specimens back to England.

In 1835, Darwin and the *Beagle* crew experienced a massive earthquake in Valdivia, Chile. The destruction was terrible. Because of the damage, Darwin observed the changes in elevation, providing evidence for theories that Earth's landmasses slowly moved over time. This experience convinced Darwin that Earth was indeed much, much older than many theologians and scholars believed at the time. In late 1835, the *Beagle* reached the Galapagos Archipelago. Here, Darwin observed that the birds varied in beak shape according to the island they inhabited. The same was true for tortoises that inhabited the various islands; the shell shapes of tortoises on different islands had different shapes.

When the *Beagle* finally returned to England in October 1836, Darwin set out to visit museums and scientific circles to discuss his specimens and discoveries. During one of these meetings, another scientist, John Gould, reviewed Darwin's notes on the Galapagos birds. He informed Darwin that there were actually twelve distinct species of finches. Over time, Darwin developed a theory that organisms best suited to their environments survive and reproduce. The traits that helped them to survive would be passed down to their offspring. This is why the Galapagos finches had different

Darwin presented the finches to the Zoological Society of London after he returned from his Galapagos voyage.

beaks. Over time, those with beaks specialized to access the primary food sources on each of the different islands, like seeds, insects, or plants, would survive and reproduce while those who were not best adapted died. The entire population would eventually evolve to have specialized beaks based on the island's food sources.

Darwin published *On the Origin of the Species* in 1859. In his book, Darwin explained his theory of natural selection: that populations change, or evolve, over time. The book also presented an argument that diverse species arose over time from common ancestors and branched off into new types of organisms. Darwin is considered the father of the theory of evolution; his theory revolutionized biology and sparked debates that continue to this day.

Throughout his career, Darwin remained passionate about plants. In the midst of writing about the diversity of living things, he also studied the effects of domestication on plants and animals. In 1876, Darwin published a book entitled *The Effects of Cross and Self Fertilisation in the Vegetable Kingdom*. The publication entailed eleven years of observation and experiments on self-pollination and cross-pollination among plants. Darwin observed that crossbreeding two different plants produced offspring superior to those produced by self-fertilization, or the pollination of a plant by its own pollen. Although the complexities of genetic inheritance were unknown at the time of his publication, Darwin drew a conclusion now supported by modern-day genetic research: cross-pollination resulted in stronger and faster-growing plants than their self-fertilized counterparts. Although he is largely remembered for his theory of natural selection, Darwin's lifelong fascination with plants would contribute to groundbreaking work by other botanists and geneticists for years to come.

Darwinians versus Mendelians

Recall that even though Gregor Mendel's work laid much of the foundation for the modern study of heredity, his experiments were largely unknown and disregarded into the twentieth century. The Dutch botanist Hugo de Vries, the German botanist Carl Correns, and the Austrian botanist Erich Tschermak von Seysenegg rediscovered Mendel's work in 1900. Each independently reported similar results on experiments with hybridization as Mendel had forty years earlier. There is mixed evidence as to whether de Vries and Correns read Mendel prior to publishing their own work as, for instance, de Vries did not refer to the monk in his original French language publication but did so in a subsequent printing in German.

Much of the popularization of Mendel's work can be credited to the English biologist William Bateson. William Bateson was one of the greatest proponents of Mendel's hereditary laws. In the early 1900s, he and his followers, called Mendelians, argued that Charles Darwin's theories on natural selection were not supported by evidence from Mendel's work and, later, their own experiments with crossbreeding. Bateson was vocal in his assertions, and he drew criticism from Darwinians, supporters of Darwin's evolutionary theory. Darwinians argued that Mendelian genetics seemed sporadic compared to the changes in traits that had been observed by Darwin and others. Other opponents believed that Mendel's theories did not apply to all organisms and were only salient for the study of plants. This division persisted for several decades until mathematical experiments demonstrated that Mendelian laws could produce large varieties of traits observed in the process of natural selection. Finally, Mendelian genetics found its place in the history of evolutionary theory.

William Bateson's contributions to the field of genetics extended beyond supporting Mendel's work. In fact, he coined the term "genetics" in a personal letter to a colleague and would later found the *Journal of Genetics* in 1910 with other visionaries in the field, such as John Punnett and Edith Saunders.

LUTHER BURBANK

The work of Mendel and Darwin inspired scientists and professionals from many different fields. Luther Burbank, a botanist and horticulturalist, is known for the development of over eight hundred varieties of fruits, grains, flowers, vegetables, and grasses. Burbank was deeply inspired by Darwin's books, especially *The Variation of Animals and Plants Under Domestication*.

Luther Burbank was born in Massachusetts in 1849, the thirteenth of eighteen children. As a child, he enjoyed helping his mother in her garden. After his father died when Burbank was twenty-one, he bought a 17-acre (6.8-hectare) plot of land in Worchester County, Massachusetts.

The Irish potato blight began four years before Burbank's birth. Caused by a fungus, the blight resulted in the deaths of many and the emigration of many of Ireland's poorest citizens, who depended on potatoes as the staple

Luther Burbank dedicated his life to developing new varieties of well-known fruits and vegetables.

of their diets. The spread of the blight did not completely eliminate susceptible potato varieties from being grown, but it did mean that potato yields were often small and sickly. Because potatoes are so versatile, many were concerned with developing a hardier version—both to avoid the consequences of the fungal blight and to ensure such a crisis of starvation and hunger did not occur again. The work of Darwin, Mendel, and Bateson, among others, was quickly changing what was known about plant breeding; however, complete understanding of heredity and plant diseases did not fully exist. Historically, potatoes were bred asexually, through self-fertilization, and breeders suspected that this practice contributed to their susceptibility to fungal blight infection. Many efforts were made to cross-fertilize potatoes, as this was believed to be key in preventing the blight;

however, these new varieties did not have the taste of the "pre-blight" potatoes.

When Luther Burbank bought his small farm, he began experimenting with plant breeding on the side, influenced by his study of Darwin's work. He began by crossbreeding different varieties of potatoes. Prior breeders had concluded that potatoes grown from seed, and not from vegetative reproduction, were resistant to the blight. When Burbank noticed a seed ball with twenty-three seeds in a potato patch, he planted each. From each seed arose a new plant. Burbank sent samples of the new variety to seed companies. Several rejected them because these potatoes had a flavor, unlike previous varieties. Finally, a company accepted his submission and named the new variety the Burbank. Later, the Burbank potato was crossed with a type of potato called a russet. Today, the Burbank russet potato is the most widely grown variety in the United States. If you have ever eaten french fries, baked potatoes, or mashed potatoes, you have likely eaten a Burbank!

This discovery cemented Luther Burbank's reputation as an expert plant breeder. In 1875, using the $150 that the seed company paid him for his potato, Burbank moved to California. There, he built and established a greenhouse, a nursery, and farms where he continued his crossbreeding experiments. Burbank rarely wrote down his procedures, instead relying on memory and acute observation for desirable traits in his plants. Despite his nontraditional record keeping, Burbank was a prolific author, writing books and plant catalogs that described all of his plants. He experimented with several breeding techniques, including grafting, cross-pollination, and hybridization.

Burbank created over eight hundred new varieties of plants, including 113 types of plums, many types of berries, potatoes, and tomatoes, among other flowers and grasses.

WILLIAM JAMES BEAL

William James Beal was another young scientist inspired by Darwin's work on fertilization. He is considered a pioneer in the development of hybridized corn. Considered one of the most successful genetically modified crops, modern corn, or maize, comes in a variety of forms including drought resistant, herbicide resistant, and insecticide producing. Today, the United States produces approximately 35 percent of the world's corn supply, and much of it is genetically modified.

First domesticated by indigenous peoples in Mexico about ten thousand years ago, corn is a diet staple across the globe. Once solely a component of early human diets, corn is now also used for animal feed, for products such as corn syrup, and as a form of fuel. While the origins of other early-domesticated plants, like rice, were relatively easy to trace, for many years the wild origins of maize were unknown. There are no wild-growing plants that resemble modern corn. Using genetic tests, scientists believe that a grass, teosinte, is corn's wild ancestor. Furthermore, scientists traced the domestication to a single geographical location in southern Mexico. Over time, early humans selected seeds from this wild-growing plant with the most desirable traits. Through planting many generations, the cobs became larger with more kernels. Corn took the form that we recognize today through the process of selective breeding. Around 2500 BCE, corn spread through Mexico and North and

Modern-day maize, or corn, does not resemble any wild-growing plant. Scientists believe that an ancient grass was the ancestor of domesticated corn.

South America. Spanish arrival to the Americas in the fifteenth century introduced corn to Europe and the rest of the world. Even though humans selected strong plants to propagate and grow, corn was extremely vulnerable to a variety of insects and diseases. Before genetic modification was a scientific possibility, a botanist named William James Beal opened the door to modifying one of humankind's oldest crops.

Beal worked as a professor of botany for forty years at Michigan Agricultural College, now Michigan State University in East Lansing, Michigan. There, he studied cross-fertilization. Specifically, Beal's research examined how to increase corn yields. His experiments are considered the first to increase the yield of maize, specifically through controlled cross-pollination. Beal's discoveries produced corn that was hardier and grew faster.

He attended Harvard University in Cambridge, Massachusetts, for his undergraduate degree. He arrived shortly after Charles Darwin's *On the Origin of the Species* was published. Darwin's theories, particularly on inheritance, inspired how Beal thought about which strains of corn to select for his experiments. Without Beal's contributions, genetically modified corn would not be possible.

In addition to his cross-fertilization studies, Beal is responsible for one of the longest standing experiments. In 1879, Beal buried twenty glass bottles, each filled with a mixture of sand and seeds on Michigan State's campus. Each bottle contained fifty seeds from twenty-one species of plants. Beal buried each bottle upside down to avoid contamination by water. Every five years, researchers were to dig up the bottles, plant the seeds, and observe which ones sprouted and which did not. Beal hoped this experiment would shed light on the vitality, or life, of seeds.

Seeds are extremely tough. They can lie dormant for long periods of time without sun or water. For example, in 2005, scientists in Israel successfully germinated, or grew, a date palm from a seed that was two thousand years old.

Caretakers continue to unearth Beal's bottles and plant the seeds today. While initially the bottles were to be unearthed every five years, researchers in the 1920s extended that to opening one every decade. Later researchers extended the time period further, to once every other decade. The last Beal bottle was exhumed in 2000. Two of the twenty-one plant species sprouted. The next bottle will be dug up in 2020 and the final in 2100.

The germination experiment and Beal's corn cross-pollination discoveries would not have been possible without the influence of early geneticists and evolutionary theorists like Charles Darwin.

DONALD F. JONES and the DOUBLE CROSS

Another pioneer in early plant hybridization experiments was Donald F. Jones, a geneticist who discovered the "double cross." Working at the Connecticut Agricultural Experiment Station in the early twentieth century, Jones contributed to the maize production industry with his breakthrough.

Despite advances in crossbreeding techniques, corn producers had not yet figured out how to combine the desirable traits from self-fertilizing corn with the positive characteristics of hybridized corn. Prior to Jones's discovery, production of hybridized corn seed at a large scale was difficult.

Jones's double cross method used four self-fertilized lines, or plants that reproduce asexually, to cross instead of the traditional two. He bred two sets of self-fertilized plants, and then he bred their offspring. The result was a hybrid plant that had more desirable traits. Such lines were developed to meet the needs of varying climates, seasons, and soil composition.

This discovery received considerable attention, and the Connecticut Agricultural Experiment Station marketed the first commercial hybrid corn in 1921. While midwestern farmers began growing hybrid maize prior to World War II, the production did not take off until after the war. From

1933, when 1 percent of United States corn was hybrid, the average rose by sixty-eight percentage points so that by 1960, 96 percent of all maize planted was hybrid. As a result, yields rose; in 1933, an average of twenty-three bushels per acre was produced; by 1980, the production rates had risen to eighty-three bushels per acre.

SIR CHARLES SAUNDERS and MARQUIS WHEAT

The twentieth century also brought improvements in the production of other global staple crops, including wheat. A type of grass, wheat was most likely first cultivated in the Fertile Crescent around 9600 BCE. Today, wheat is grown and consumed globally. In 2017, 740 million metric tons (815.7 million tons) were consumed across the world. Thirty million metric tons (33 million tons) were eaten in the United States. One of the more famous types of wheat, marquis wheat was developed in Canada at the turn of the century. Although wheat was first planted in Canada in 1605, many of the varieties grown were not well-suited to the weather conditions found throughout much of the country. In 1842, the introduction of a new strain, red fife, contributed to increased production, thereby enabling Canada to expand its trading system, including railways. Sir Charles Edward Saunders was a Canadian agronomist. Agronomy is the science of producing plants for human use. It combines the disciplines of biology, genetics, chemistry, ecology, and economics. Saunders's father, Dr. William Saunders, appointed him as an experimentalist at the Experimental Farm in 1903. Saunders observed

Marquis wheat is the most widely grown variety across the globe.

that despite its popularity, red fife was susceptible to late fall frosts common in western Canada. He believed that the introduction of wheat from other parts of the world could improve red fife's growing ability. Saunders set about developing a superior strain of wheat using the plant breeding techniques, like cross-fertilization, available to him at the time. He was particularly interested in wheat that would produce the largest loaf of bread when baked. To accomplish this, Saunders first selected superior plants. He would then actually chew kernels from these plants to determine which formed the best dough ball. A larger dough ball, he surmised, indicated that the plants had a superior gluten concentration and would thus produce the largest loaves of bread. He also tested his theory by actually baking flour produced from different wheat strains. His final selection, marquis wheat, matured faster than red fife and had impressive yields at 41.6 bushels an acre. It also retained the high bake quality that made red fife popular. Marquis was a cross between Canadian and Indian wheat varieties. For over fifty years, marquis was the dominant strain of wheat grown around the world.

Each of the discoveries discussed in this chapter was fundamental to the fields of botany and agronomy. Through this body of research, yields increased and plant hardiness improved. However, combining these techniques with discoveries about heredity by early geneticists like Crick, Watson, and Nirenberg would enable scientists to produce organisms never seen before: genetically modified crops.

The Burbank Potato

A 1920 newspaper article from the *Stockton Daily Evening Record* tells the story of the Burbank potato. In his article, entitled "Luther Burbank Tells the Origin of the Burbank Potato," the famous plant breeder recounts the story of how he developed the most prolific potato type today:

> I had been raising seedling potatoes, but they all came out, unfortunately, like the parent plant, and were not good. Just then I came across this seed ball. It was a remarkable experience. I felt that something remarkable would develop. In the fall of 1872 I planted the seeds out in a field. I used to visit every day, but one day I lost it. I felt concerned and hunted day after day, and at last I found it about twenty feet from where I had planted it. It is remarkable that such a development came from such an incident as that ...

The development of the Burbank potato was the result of careful and calculated cross-fertilization experiments, much like Darwin's observations of plant fertilization years before. He goes on to describe the initial skepticism from the first seed company he offered his findings to:

> I wrote to B. K. Bliss & Co. of Rochester, N. Y. in the spring of 1873. I sent a sack. I felt glad that I had a wonderful potato. The firm replied that they were so sweet and nutty that they feared they must be frozen, and, therefore turned them down. I sent a sample to James H. Gregory, Marblehead, Mass. And was asked to see him. I went. He asked what I wanted for them. I asked $500. He offered me $125 ... now they are so plentiful that since then enough of them have been grown that they would require a freight train 14,000 miles long to carry them, or in other words, the train would be long enough to reach from Santa Rosa to any point on this planet.

Burbank's cross-pollination experiments addressed many modern agriculture concerns. Avoiding crop loss while producing enough food is one of the main catalysts for the development of genetically modified crops.

The Flavr Savr tomato was the first genetically modified crop to be sold at large scale.

CHAPTER 4

The Discovery of Genetically Modified Crops

The nineteenth and twentieth centuries brought changes in our understanding of how traits are passed down from one generation to the next. Dr. Marshall Nirenberg, for example, discovered the central dogma, or the transfer of DNA instructions to RNA, which is then translated into proteins. Scientists learned how to breed hardier and better-adapted plants. They discovered the genomic sequences of many living things. Breakthroughs like recombinant DNA set the stage.

GENETIC DISCOVERIES

Recall that genetic engineering includes the modification of an organism's genes by the manual insertion or alteration of its genome. Genetic modification produces organisms with more favorable traits like survival in adverse environmental conditions, resistance to the effects of chemical pesticides and herbicides, and the production of higher yields.

Early attempts to genetically modify crops were driven by the need for crops that would not die at low temperatures. In the 1960s, researchers at the United States Department of Agriculture (USDA) observed that some plants in an experiment died because of an unexpected frost while others did not. The cause, they discovered, was actually bacteria that lived on the surface of the plants. These bacteria produced a protein that ice could collect on and ultimately kill the plant. However, due to a mutation, some bacteria did not produce the protein, and thus, plants with these mutated bacteria were able to survive the colder temperatures. This discovery was exciting; the potential to produce plants with

The herbicide glyphosate was commercially marketed beginning in 1974.

only the mutated bacteria could save millions of dollars in crops lost due to weather. A number of companies investigated producing a genetically modified form of the bacteria that would never produce the ice proteins. In 1983, Advanced Genetic Sciences applied for a permit from the USDA to test a spray, Frostban, that would coat plants with genetically engineered bacteria. The USDA granted the permit, the first given for testing a genetically modified crop. The Frostban tests were delayed, however, due to protests. Advanced Genetic Sciences did not pursue commercial production of Frostban.

Another attempt was the Flavr Savr tomato. The Flavr Savr was intended to ripen more slowly than naturally occurring plants. Slow-ripening tomatoes would save millions of dollars in fruit that rotted too quickly and had to be thrown away before even being sold. An enzyme, or special protein, is responsible for the ripening process in tomatoes. This enzyme attacks the fruit's skin, causing it to break down and become soft. Researchers theorized that deactivating the enzyme could slow the ripening process.

The company submitted a request to the Food and Drug Administration (FDA), and the FDA tested the Flavr Savr for several years. The FDA concluded that the genetically engineered tomatoes were safe for consumption, and, for some time, they even outsold their "natural" counterparts in supermarkets. However, sales slowly began to drop off, and, by 1999, the Flavr Savr tomato was pulled from the shelves.

Other early genetic modification efforts included the development of the herbicide Roundup, as well as the subsequent crops that were modified to resist the effects of that herbicide. Monsanto, one of the largest agrochemical companies in the world, discovered that the chemical glyphosate is highly effective at killing weeds. Marketed as Roundup, the herbicide soon became extremely popular among farmers. In fact, a study by the United States Department of Agriculture found that the usage of Roundup and similar herbicides grew from basically zero in 1990 to ten million pounds (4,535,923.7 kilograms) in six years. However, Roundup was so effective that it would sometimes kill off the crops that it was meant to protect. In response, Monsanto created plants that were genetically modified to resist the effects of Roundup. It did this by inserting a gene into the plants that blocked the effects of glyphosate. Approximately 14 percent of corn, 15 percent of cotton, and 93 percent of soybeans grown today are genetically modified to be Roundup resistant.

Early discoveries like Roundup-resistant plants, the Flavr Savr tomato, and cold weather plants drove the development of genetically modified crops today. Let's examine the pioneers that paved the way for modern plant genetic engineering.

HERBERT BOYER and STANLEY COHEN

Herbert Boyer was born in 1937, in Darry, Pennsylvania. After graduating with a bachelor's degree in chemistry and biology, Boyer pursed a doctorate degree in Pittsburgh, Pennsylvania. After receiving his degree, Boyer did his post-graduate work at Yale University in New Haven, Connecticut. Stanley Cohen was born in 1922, and he also majored in biology and chemistry, at Brooklyn College in Brooklyn, New York. He graduated with his doctorate

Herbert Boyer (*left*) and Stanley Cohen (*middle*) have both been honored with the Biotechnology Heritage Award for their invention of recombinant DNA technology.

degree in biochemistry from the University of Michigan, in Ann Arbor. The two scientists did not meet until 1972, when they were both at a conference in Hawaii. At the time, Cohen was a professor at Stanford University in Stanford, California, and Boyer was teaching biochemistry and performing research at the University of California in San Francisco. After meeting at the conference, the two met at a deli to discuss their respective research and found they had similar interests. After that meeting, the two formed a partnership to investigate cloning DNA segments. Boyer and Cohen developed a DNA copying process called recombinant DNA.

They discovered that, by using a bacterium, they could copy segments of DNA quickly and cheaply. Recombinant DNA could be used to combine the DNA from two different organisms. Recombinant DNA requires a cloning vector. Typically, vectors contain signals for DNA replication that will reproduce combinations of DNA. A fast and inexpensive method of copying DNA was a key discovery in the mass production of genetically modified crops.

Designing a Genetically Modified Plant

So how is a genetically modified plant actually made? There are several commonly used editing techniques. One of the early genetic engineering techniques is the gene gun. Monsanto used the gene gun technique to create the Roundup-resistant soybean. Gene guns operate by literally shooting DNA into the nucleus of plant cells. Particles coated with genetic information are injected. Once inside the nucleus, the genes can potentially insert themselves into the DNA of the cell. The gene gun method, however, is not foolproof. Often, too many DNA copies can be inserted, and the cell will detect a problem. In this case, the foreign DNA is not replicated by the cell.

Because success is not guaranteed with the gene gun method, scientists created other strategies to get DNA from other organisms into the cells of plants. Another genetic engineering process utilizes a naturally occurring soil bacterium as a sort of genetic engineer. In the wild, *Agrobacterium tumefaciens* invades plants through wounds in their root systems or stems. Once inside, part of the bacterium's DNA inserts itself into the plant's DNA. The

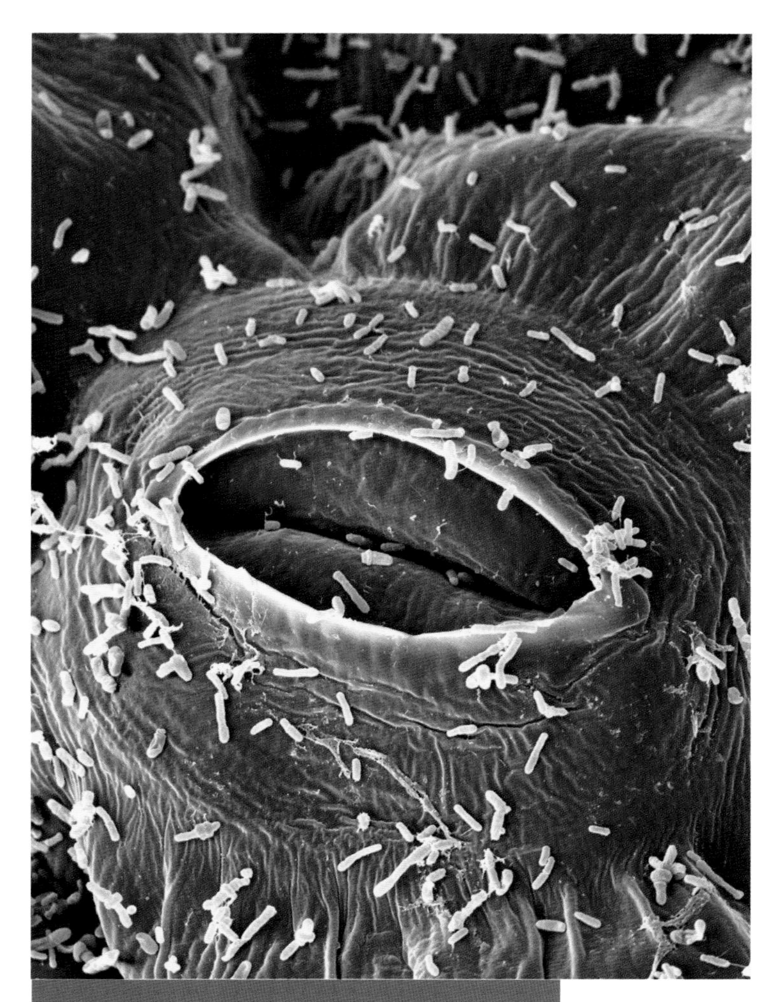

Agrobacterium tumefaciens is used in agricultural research as a carrier to introduce genes into plant tissues.

pieces of bacteria DNA then force the plant to produce materials that *A. tumefaciens* needs to survive. The plant will continue producing these materials until it dies. At that time, the bacterium enters the soil until it encounters another wounded plant.

Geneticists adapted the *Agrobacterium tumefaciens*'s natural genetic engineering techniques to insert other genes into plants. Instead of the harmful instructions that kill a plant, scientists insert helpful genes into the bacterium, and it then delivers that genetic information to plant DNA.

A third way that scientists genetically engineer plants is through the use of microfibers. Microfibers act much in the same way gene guns do. The microfibers are coated in desired genes. Also called "whiskers," the microfibers are placed in a test tube with cells. The whiskers puncture the cells, delivering the genetic material to the cellular nuclei without damaging the cells.

There are other methods that do not pose any risks to a cell's nucleus. For instance, using pulses, the process of electroporation opens the pores in plant cells and allows the delivery of genetic material to the cell.

A more complex method of genetic alternation is incorporating stacked genes into an organism's DNA. All the methods previously discussed are for inserting a singular gene change. For instance, scientists may use *Agrobacterium tumefaciens* to deliver insect-resistant genes to a plant's genome. What if scientists wanted to produce a plant that had multiple genetic modifications, like faster growth and insect resistance? Genetic engineering techniques like gene guns and electroporation take some time and are difficult technical procedures. Scientists developed a solution to these problems by creating "stacked" genes. Imagine two genetically modified plants of the same species. One has been modified with insect-resistance genes and the other with genes to make it grow faster. By cross-pollinating these two plants, scientists increase the chances of producing offspring with both desired genes. Through breeding subsequent generations, scientists can add even more additional modifications. Stacking genes is the process of adding genetically modified traits to an organism's genome by breeding multiple generations. The agrochemical company Monsanto and the Dow Chemical Company have created corn with eight stacked genes. These genes include resistance to several types of insects and herbicides.

PLANT GENETIC SYSTEMS

One of the leaders in plant engineering is the company Plant Genetic Systems. Founded in 1982, the company specializes in the *Agrobacterium tumefaciens* vector method of genetic modification. The two founders, Marc Van Montagu and Jeff Schnell, both worked at the University of Ghent in Belgium.

Marc Van Montagu was born in Ghent, Belgium, on November 10, 1933. At the age of twenty-two, he graduated with his master of science in organic chemistry and subsequently received his doctorate in chemistry ten years later. He would divide his career among several pursuits: as a full-time professor at the University of Ghent, as a part-time professor at the Free University of Brussels, in Belgium, and as the director of genetics at the Flanders Institute for Biotechnology, also in Belgium. As a researcher, Van Montagu and his team discovered the genetic engineering properties of *A. tumefaciens* bacteria. Specifically, they found the mechanism that the bacteria used to change plant DNA. This molecule was called a Ti plasmid. A plasmid is a small DNA molecule within a cell that is independent of DNA contained in chromosomes. Plasmids can replicate separately from the rest of the cell's DNA. The Ti plasmid in *A. tumefaciens* is made up of 196 genes. Van Montagu and other researchers discovered that the Ti plasmid is the tool by which the invasive bacteria alters wounded plants. This discovery helped researchers understand how to use *Agrobacterium tumefaciens* to deliver genes in order to grow genetically modified plants.

The second founder of Plant Genetic Systems (located in Ghent), Jeff Schnell studied zoology and microbiology in Belgium. Schnell's work primarily focused on the interactions between plants and soil bacteria—a perfect complement to Van Montagu's bacterial research!

In 1985, Schnell and Boyer's company announced that they had designed tobacco plants that were resistant to disease-causing insects. They did so through the use of soil bacteria, *Bacillus thuringiensis* (Bt). These bacteria live in many environments, including the stomach of some caterpillars, in water environments, on the surfaces of some plants, and in animal droppings. Bt produce a protein that is toxic to many different types of insects. Called crystal proteins, or cry toxins, these proteins are lethal to a variety of insect species that damage and cause disease in plants, including moths, butterflies, flies, mosquitos, beetles, ants, wasps, bees, and nematodes. When an insect ingests the bacteria and the crystal proteins, the cry toxins paralyze the digestive tract. The insect stops eating and eventually starves to death. Many Bt bacteria may colonize, or inhabit, the digestive system of the insect, and this can also result in death.

Researchers at Plant Genetic Systems designed plants with genes for Bt production. With the crystal proteins, the crops were resistant to disease introduced by insect pests. With insect-resistant plants, farmers did not have to apply as much insecticide to their crops. Bt technology quickly became very popular, and a variety of Bt modified crops were produced, including cotton and tomatoes.

Supporters of Bt-modified crops argue that they have several benefits. First, they require farmers to use fewer insecticides, and thus they have a positive effect on the environment. Supporters also point to the economic benefits of Bt crops for farmers as they save money by not buying expensive insecticides.

Evidence of these benefits is mixed. Research on Bt crops has also found moderate evidence of increases in yield. A study of Bt cotton in India found that, when compared to non-Bt cotton, the Bt yields were higher by 80 percent. A different study of Bt cotton production in Arizona found much smaller yield increases, about 5 percent. Lastly, a seven-year study of Bt cotton in China found that, initially, insecticide production was reduced and farmers saved money. This reduced use, however, did not last, as other insects damaged plants. Farmers subsequently increased their use of insecticides once again.

2013 FOOD PRIZE WINNERS

At the same time discoveries were made at Plant Genetic Systems, another scientist, Mary-Dell Chilton, was researching similar aspects of crop modification. Mary-Dell Chilton was born in Indianapolis, Indiana, in 1939. She graduated with a doctorate degree in chemistry from the University of Illinois at Urbana-Champaign. This was at a time when fewer women attended college than in modern times, and even fewer women pursued degrees in science, engineering, and math.

Chilton performed postdoctoral research at the University of Washington in Seattle. In 1977, she led a team to discover that plant-harming genes could be removed from the *Agrobacterium* while retaining its ability to insert its DNA into a plant's cells. While Van Montagu and his team discovered the Ti plasmid, it is Chilton's work that actually determined how to disarm and use the plasmid to insert desired genes into a plant's genome. Chilton and her team at Washington University in St. Louis are credited with creating the first *Agrobacterium*-modified transgenic plants. Because of this discovery and other genetic engineering feats, Chilton has been called the "queen of *Agrobacterium*."

In later interviews, Chilton recalled that it was DNA that initiated her fascination with genetic engineering. Until Van Montagu's discovery of *Agrobacterium*'s genetic engineering potential, she did not believe it was possible because plants and bacteria were so different. In 1983, she began working for an agrochemical company now called Syngenta. Her research continued to contribute to the development of disease- and insect-resistant, environmentally hardy plants.

Another genetic engineering pioneer, Robert Fraley grew up on a small farm in Illinois. He studied microbiology and biochemistry and, after receiving his doctorate degree in 1978, spent a year doing postdoctoral research at the University of California, in San Francisco. After completing his fellowship, Fraley was hired by the agribusiness Monsanto, where he worked his way from senior research specialist to executive vice president and chief technology officer.

In 1984, Fraley worked on the team responsible for the design of the Flavr Savr. He also contributed to the development of the Roundup-resistant plants. Fraley and the Monsanto research team also published a paper on the first genetically modified crop, mentioned in chapter 2, a tobacco plant designed to be antibiotic resistant.

Tobacco is not the only plant that contains nicotine, a naturally occurring substance that wards off insects, but humans have used it for many thousands of years for nicotine's stimulant properties. Around six million tons (5.4 million metric tons) of tobacco are produced each year, with a large percentage for cigarettes. The World Health Organization (WHO) has named tobacco use as the single most preventable cause of death globally.

Today, tobacco is one of the most popular plants to use for genetic engineering experiments. Discoveries that were made using tobacco include the insertion of human genes to produce antibodies, modified tobacco plants that can clean the soil, and the possibility of using tobacco as a biofuel.

In 2013, Marc Van Montagu, Mary-Dell Chilton, and Robert Fraley were awarded the World Food Prize.

The World Food Prize is one of the most prestigious international recognitions of innovation and discovery in the pursuit of the alleviation of global hunger, specifically the quality, quantity, or availability of food. An annual prize of $250,000 is given to the recipients. In addition to their contributions to genetic science, the prize recipients share defining experiences with food and agriculture. Van Montagu, for instance, grew up in Europe during World War II when food rationing and scarcity were common. Fraley's childhood experiences on his family's farm influenced his desire to develop better agricultural tools.

BARBARA MCCLINTOCK

The year 1983 was quite an exciting one for the discovery of genetically modified crops and the world of genetic engineering in general. First, Kary Mullis developed PCR, polymerase chain reaction. Van Montagu and Schell's Ghent Laboratory discovered the Ti plasmid, and Mary-Dell Chilton's team created the first *Agrobacterium*-transgenic plants. Robert Fraley and the Monsanto team developed genetically altered Roundup-resistant plants.

Barbara McClintock's research also contributed to this stellar year for plant engineering discovery. McClintock's study of maize chromosomes contributed not only to the development of genetically modified crops, but also to our greater understanding of chromosomal inheritance.

Born in 1902, McClintock graduated with a doctorate in botany from Cornell University in Ithaca, New York, in 1927. She studied mutations in maize and discovered transposable, or jumping, genes. Though originally her findings were not widely acknowledged in the scientific community, she is now considered one of the foremost cytogeneticists in the world. Cytogenetics is the study of chromosomal behavior, particularly during cellular reproduction.

Transposable genes move around, or jump, from one location to another. When McClintock made this discovery, she proposed that such DNA sequences could play a supervisory role in the cell, determining which genes are turned on or off. Initially, other scientists dismissed transposable genes as "junk" DNA. However, now scientists believe that jumping genes may make up at least 40 percent of all human DNA. Later experiments would also confirm the influence of transposable elements on the cellular reproduction of maize, solidifying McClintock's earlier hypotheses.

McClintock received over thirty-one prestigious awards and honorary doctorates for her discoveries, including the Nobel Prize in Physiology or Medicine in 1983. The first woman to win the prize unshared, it was awarded for her discovery of transposable elements, which the Royal Swedish Academy of Sciences compared to Mendel's genetic discoveries. Better understanding the genome of maize was an important step in developing genetically modified corn, one of today's most common genetically modified crops.

MONSANTO

The final GMC innovator is not an individual, but a company. Monsanto's creation of Roundup and Roundup-resistant plants is not the only contribution this mega agribusiness has made. One of the most well-known and

Monsanto is often at the center of public discussions on genetically modified crops.

controversial modern agricultural corporations, Monsanto produces many of the world's genetically engineered seeds.

Monsanto was originally founded in 1901 to produce the sugar substitute saccharin. The founder, John F. Queeny, named the company after his wife, Olga Monsanto Queeny. To start the company, Queeny invested $1,500 of his own money and borrowed another $3,500 from an Epsom salt producer. By 1902, the company added caffeine and vanillin to its full-scale saccharin operation. By 1915, with the introduction of Coca-Cola as a major customer of Monsanto, the company had over $1 million in profits. Two years later, aspirin was added to the production line.

Both World War I and II were extremely beneficial for Monsanto. The company expanded during the First World

War. After John F. Queeny passed the ownership to his son, Edgar M. Queeny, the company was incorporated as the Monsanto Chemical Company, in 1933. It produced styrene, a component of synthetic rubber, which was a critical component of the United States' war efforts.

Over the next six decades, the company focused on the production of chemicals like sulfuric acid. In the 1960s, Monsanto formed an agricultural division to focus on the production of herbicides. In 1964, the name was changed to Monsanto Company.

The company became one of the primary suppliers of Agent Orange during the Vietnam War. The United States sprayed the herbicide on Vietnamese jungles and fields to kill food sources and destroy foliage cover. Agent Orange has been found to have damaging environmental and health consequences. Reforestation in Vietnam was slowed or made impossible after the war. The 20 million gallons (75.5 million liters) of Agent Orange and other herbicides sprayed also affected wildlife. Agent Orange is also harmful to humans. Researchers have found an increased incidence of cancer and respiratory disease in both Vietnamese civilians and American veterans of the Vietnam War.

In 1976, Monsanto created Roundup. Recall that the herbicide became widely sold and globally popular. In the 1990s, Monsanto acquired other biotechnology companies, which enabled it to expand in its influence on the development and production of genetically engineered seeds and plants.

In 1996, the company introduced the first biotech crop Roundup Ready soybeans, designed to withstand the effects of Roundup herbicide. At the same time, Monsanto released cotton genetically engineered to resist pests and insects. Two years later, Roundup Ready corn was introduced.

Throughout Monsanto's existence, critics have voiced concerns over the company's ecological, health, and economic impacts. A full discussion of the controversy surrounding the agribusiness giant will be presented in the final chapter.

BENEFITS of GENETIC MODIFICATIONS

Today, genetically modified crops are engineered to have a diverse and varied set of traits, uses, and benefits. Genetic modifications include disease, pest, and herbicide resistance; environmental stress resistance; improved shelf life; improved nutritional value; increases in yield; remediation of environments; and other products for medicine and transportation.

Many scientists also see the potential for genetically modified crops to improve world hunger rates and provide vital nutrition, particularly in developing countries. For instance, the International Rice Research Institute (IRRI) has developed golden rice that has higher quantities of vitamin A.

Golden rice can help those with vitamin A deficiencies, most common in Africa and southeast Asia. Vitamin A deficiency results in extremely negative effects, including blindness, in young children and pregnant women in low-income countries.

Other genetic modifications include the production of medicine or biological substances. For instance, a version of genetically modified tobacco produces human antibodies, or types of cells used to counteract pathogens. Other research seeks to convert plants, like algae, into biofuel to be used in vehicles.

Finally, bioremediation refers to plants engineered to remove the effects of pollution and contamination from ecosystems. To create such plants, scientists insert genes from bacteria into a weed that allow it to remove harmful chemicals from the soil.

Today's genetic engineering research focuses on improving crops, particularly in developing nations. Improvements include cereal crops, such as wheat or grain that can "fix," or convert, nitrogen into helpful molecules for them. This is normally a process carried out by bacteria in nature or in commercial agriculture by fertilizer.

Self-nitrogen-fixing plants could save farmers, particularly those in developing countries, money that would normally have to be spent on expensive fertilizers. Research also must take into account changing temperatures that accompany ongoing climate change, like developing tomatoes that ripen earlier to account for warming temperatures.

cDNA libraries

Delivering genes using a bacterium vector can be a complicated process. Luckily, there are some methods that ensure that the DNA being delivered is as straightforward and easy to use as possible.

Recall that recombinant DNA combines the genetic material from two different organisms. Often, a large number of clones, or copies, of the DNA sequence form what is called a DNA library. DNA libraries are categorized by what vector is used to make the copies and the size of the genome that is being copied.

A cDNA library differs from others because it utilizes another type of genetic information holder, mRNA, to create a single-stranded DNA copy. Because the copies come from the mRNA template, they do not have introns, or sections of DNA, interrupt the order of genes. Because DNA copied from the mRNA does not have these stopping and starting points, the DNA from eukaryotes can be implanted into prokaryotes, or bacteria. These bacteria can then express eukaryotic genes.

cDNA libraries are often used when making copies of eukaryotic genomes because they remove the non-coding regions. cDNA can also be inserted into a vector, like the Ti plasmid, to be delivered

Scientists need to create many DNA copies when studying genetic engineering techniques.

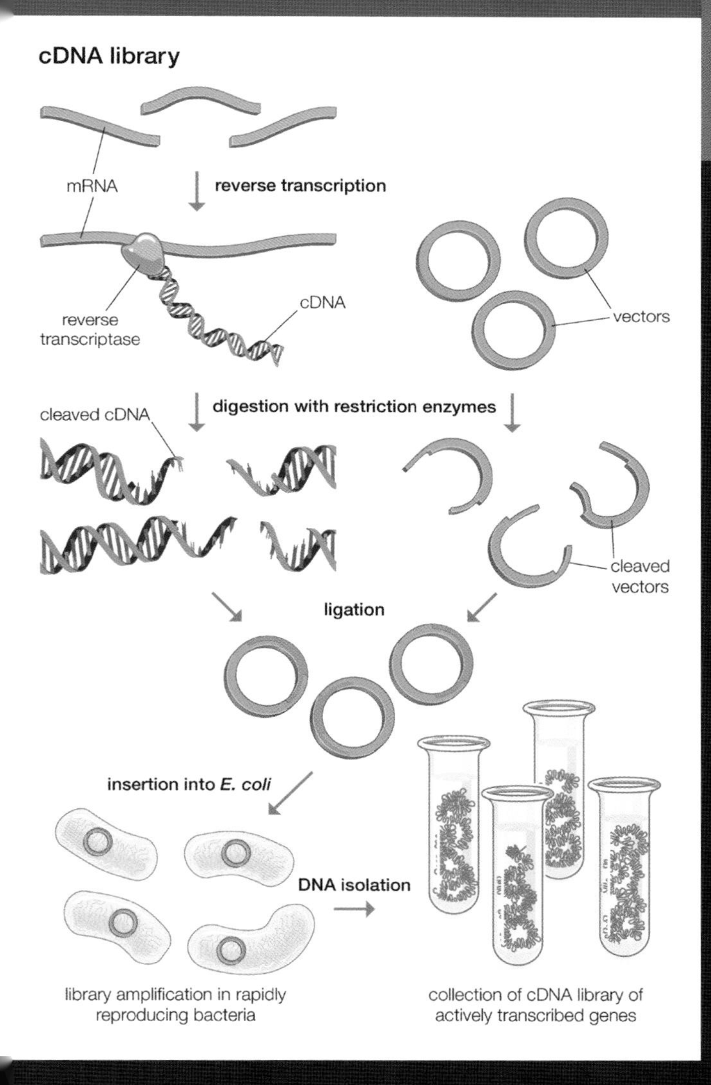

nto a different organism. cDNA libraries can also be helpful in identifying desired genes as it eliminates extra material.

Soybeans are the second most planted crop in the United States after corn.

CHAPTER 5

Genetically Modified Crops Today

Today, over 90 percent of soybean and corn grown in the United States is genetically modified. The bulk of genetically modified corn is used to feed livestock. Other uses include the production of high-fructose corn syrup used in many processed, packaged foods and corn starch. Biofuel is also made from corn. Biofuel is produced through biological processes, like agriculture. It differs from traditional fuel, like coal and petroleum, which come from fossil fuels. (Fossil fuels are composed of ancient organic material.) Genetically modified soybeans are primarily grown to produce soybean oil, a staple in many restaurant kitchens. Another widely grown genetically modified crop is cotton. It is grown to produce cottonseed oil, which is used for making processed foods and frying things like chips and margarine.

MODERN CRISES

While transgenic crops have existed since the early 1980s, there are new potential uses for genetically engineered crops on the horizon as global populations increase and as Earth's climate changes.

There are more than 820 million people in the world who are hungry. Feeding an ever-growing global population is one of the most pressing agricultural dilemmas of today. Even in the United States, approximately one in seven families do not have reliable access to nutritious food. Many of the natural resources needed to produce enough food, like water and land, have been depleted through non-sustainable agricultural practices. Producing food for an ever-growing world population while responsibly using the finite resources available is a global challenge.

Often, governments regulate agriculture. For instance, farmers sometimes face conditions out of their control that can affect their crop yields, thereby impacting their livelihoods. The United States provides income support to farmers and assistance in the event of disasters, like floods and fires. In addition to government programs, the developments in agriculture over the past century, including genetic modification, have been shared with developing countries. International programs and foundations introduced GMCs and dramatically increased production. Such programs operate on the belief that with more resilient, higher-producing crops, farmers in developing countries can grow more for less money and feed more people. By 2010, GMCs were grown in twenty-nine countries. Other supporters of GMCs argue that with climate change affecting food outputs, these hardier crops must be embraced.

Marc Van Montagu, one of the 2013 World Food Prize recipients, wrote an article praising genetically modified crops. He points out that by the early 2000s, 9 percent of all agricultural land was growing genetically modified crops. He also explains the research evidence that GMCs produce higher yields at lower costs to the environment. Moreover, Van Montagu writes hopefully of the potential for GMCs in hunger alleviation and support of developing countries.

> These crops will continue to reduce hunger by bringing more bountiful and nutritious harvests. They will also help the environment by mitigating the impact of agriculture by conserving our precious, finite supply of fresh water; freeing up land for other uses, like carbon-absorbing forests; preserving topsoil; and reducing the use of insecticides and herbicides, thereby enhancing biodiversity. … The question of how to nourish two billion more people in a changing climate will prove one of the greatest challenges in human history. To meet it, we should embrace an agricultural approach that combines the best features of traditional farming with the latest technology.

Van Montagu also articulates the controversy surrounding genetically modified plants—in particular, the concerns surrounding superweeds, the safety of consuming genetically modified crops, and the devastating effects on the environment. The following sections examine the controversies and criticisms surrounding GMCs.

CONTROVERSIES

Just as when Rachel Carson published her book *Silent Spring* on the destructive effects of the pesticide DDT, criticisms against the potentially harmful consequences of genetic modification have existed since the early 1980s. Current research indicates that consumption of genetically modified foods is not harmful; however, critics argue that there may be long-term effects that are currently undetectable. They contend that because of the relatively short time that GMCs have been sold to consumers, trends may not be visible yet. Other critics worry that GMCs are unsafe to eat; however, those claims are largely unsupported by scientific evidence.

Labeling Genetically Modified Foods

One of the concerns surrounding the sale of genetically modified crops to the public is food labeling. By 2002, over 60 percent of processed foods sold in the United States contained ingredients from genetically modified crops. By 2010, approximately 75 percent of processed food sold contained GMCs. Critics of GMCs maintain that the public has a right to know if the produce and processed foods that they are buying have been genetically altered. Supporters of GMCs, on the other hand, contend that labeling is unnecessary and will only negatively impact the sales of those products. Currently, the Food and Drug

Protestors march in support of labeling foods containing genetically modified ingredients.

Administration requires genetically modified products to be labeled only if the modifications have significantly changed the food item from its naturally occurring counterpart. Labeling is also required if the item poses a significant safety concern. Currently, there are no federal laws that require labeling in the United States; however, some states have passed their own laws. In 2016, Vermont became the first state to require labeling for all foods containing genetically modified components.

Like states, some companies have taken matters into their own hands. Whole Foods, the organic grocery store, for instance, has pledged to label all products that contain genetically modified components.

Policies

Globally, there are more labeling regulations; for instance, the European Union (EU), a collective governing body comprised of twenty-eight European countries, has passed stricter labeling, growing, and trading laws around GMCs. The United States government has a fairly relaxed attitude toward GMCs. United States policy focuses on the final food product rather than production processes. The FDA classifies genetically modified crops as generally accepted to be safe. As long as companies can prove that their GMCs are not extremely different than non-modified organisms, they are cleared for marketing and sale. The USDA requires that companies submit data on new GMCs before they can be introduced in America.

On the other hand, the European Union focuses more intensely on the production processes. The EU decided its GMC policy in 2002. In the European Union, all genetically modified crops must go through a premarket approval process and must follow strict labeling requirements. Any food that contains .9 percent or greater genetically modified components must be labeled as such. Companies must submit applications anytime they wish to introduce a new genetically modified crop. The European Food Security Authority (EFSA) assesses the product and application. If the EFSA approves an application, representatives from the EU vote. Even if the EU approves a GMC, the decisions of each individual country take precedent; however, the EU can find opposing decisions unlawful.

Because of these two governing bodies' differing approaches to regulation, the United States has approved a higher number of GMCs for growing and consumption than in Europe. These different policy approaches also contribute to differences in land use and cultivation; in 2013, the United States cultivated approximately 70 million hectares (172.9 million acres) of genetically modified crops, whereas the twenty-eight European Union member states collectively grew less than .1 million hectares (about 247,000 acres). Despite these differences, both governing bodies adhere to strict scientific assessment guidelines of the potential health and environmental effects of new crops.

Monsanto

The agrochemical company Monsanto sits at the center of much of the controversy surrounding GMCs both in the United States and abroad. Several documentaries have been made that criticize the company for its past history of producing chemical weapons and for its corporate monopoly on genetically modified seed production. Again, proponents of Monsanto's agricultural productions argue that enough research points to the safety of genetically modified plants while critics argue that there is not enough evidence that GMCs are safe.

Despite Monsanto's obvious economic success, the company remains contentious in the public eye. Lawsuits and public protests have both been leveraged against the

March Against Monsanto is a political movement founded in 2013.

company. Monsanto has also been the plaintiff in several high-profile instances of litigation, most often to protect its patents, or ownership, of genetically modified seeds.

One of the major critics of Monsanto is the activist group March Against Monsanto. The group promotes buying organic foods and boycotting Monsanto and its subcompanies. The group hosts peaceful protest marches annually in over three hundred cities. The first march took place in 2013.

In the midst of the controversy and consumer resistance to purchasing genetically modified crops, Monsanto still employs ancient crossbreeding techniques to identify desirable traits and produce new versions of plants that have a mixture of traits in addition to high-tech genetic engineering production.

Intellectual Property

Supporters of genetic modification technology argue that new engineering processes hold the key to addressing global hunger and environmental damages. Lawmakers and GMC production companies like Monsanto must figure out how to regulate ownership of these technologies. Part of the conflict comes from balancing the rights of farmers and the rights of the GMC designer. Seed copyrights, or intellectual property, allow whoever created a genetically engineered plant to retain all ownership of technology and production. Patent law means that the owner has all rights to the sale and production of genetically modified seeds. However, farmers have long engaged in the practice of seed saving, or keeping seeds for the next growing season. Seed saving, however, violates patent law.

Attempting to maintain a fair balance has even lead to lawsuits. In 1980, the Supreme Court decided that patents could be awarded to living things, like GMCs. Critics argue that patents on genetically modified plants prevent others, particularly research universities, from doing research and making discoveries. There is also worry that this allows patent holders to charge prices for seeds that are too expensive for farmers to afford. Proponents point out that, because genetically modified seeds often have advantages in terms of growth and hardiness, the higher seed costs may be lessened by the GMC's growing ability. Genetically modified plant patents last for twenty years. The patent on Monsanto's Roundup Ready soybeans invented in 1996, for instance, expired before 2016. After the expiration of a patent, the plant becomes public knowledge to which other companies and research institutions have access.

Monoculture and Superweeds

Although a diverse number of species are available for human consumption, a strikingly low number of crops are actually grown in the United States. This practice is referred to as monoculture, and it is another issue facing modern agriculture. Monoculture is the agricultural practice of growing the same or similar crops in the same area over a long period of time. This is the predominant agricultural practice in modern-day America. While much of the monoculture crops grown today are genetically modified,

Lolium rigidum is the first discovered superweed.

having little variation among crops makes them susceptible to disease and pests.

Recall that glyphosate is the chemical discovered and marketed by Monsanto as Roundup pesticide. Glyphosate is easy to work with and less harmful to the environment than its predecessors, and it breaks down quickly. After the development of the pesticide, Monsanto created crops that were resistant to the chemical, meaning that harmful weeds could be destroyed without harming the plants. However, over time, glyphosate-resistant weeds appeared. Because farmers sprayed so much Roundup, the Darwinian principle of evolution, or change over time, happened very quickly. Weeds that were not killed could reproduce and pass on their resistance genes. Called "superweeds," these plants can carry out the life processes typically disabled by Roundup. The first resistant weed, *Lolium rigidum*, a grass species, was discovered in an apple orchard in Australia in 1996. In 2000, superweeds were spotted in the United States, in a Delaware soybean field. As of 2017, scientists have found thirty-eight different species of glyphosate-resistant weeds.

Because of the evolution of Roundup-resistant plants, farmers are being forced to resort to past methods of weed management, including hand pulling and plowing. Scientists and farmers worry that if this problem is not addressed, it could limit agricultural productivity. Further, as more species become resistant to herbicides, farmers must use more herbicides. Scientists must figure out ways to make crops also more resistant to herbicides.

Some critics argue that genetically modified crops may also have negative effects on biodiversity. Biodiversity refers to variation among living things in the natural world. Biodiversity among agricultural commodities is particularly important as a greater variety of genetic information ensures the resilience of crops against many different environmental conditions. As climate change affects the conditions in which plants grow, this type of resilience is critical to sustaining large enough yields to feed the growing global population. When GMCs were introduced, there were concerns about growing large amounts from a small genetic pool. However, research indicates that glyphosate-resistant plants have not had a significant effect on crop diversity. Studies do show, however, that Bt-modified crops initially did reduce the genetic diversity of harvests. On the other hand, genetically modified crops could actually improve diversity with the introduction of undergrown plant varieties. There are an estimated seven thousand species of plants available for human consumption. Even so, only four crops (wheat, maize, potatoes, and rice) comprise over one-half of the total global agriculture output. Introducing different varieties through genetic modification could help diversification of global crops.

Do GMCs influence the biodiversity of ecosystems? Other research has found little effect of Bt-modified crops on soil organisms like worms and mites. Evidence does show a negative impact of Bt maize on soil, reducing soil respiration, or the production of carbon dioxide by plants.

Another environmental concern is that GMCs may impact the loss of natural land for agriculture growth. Particularly since the inception of large-scale industrial agriculture, environmentalists have worried about the loss of forest, wetlands, and other ecosystems to growing farms. Genetically modified crops may play both a positive and a negative role in land loss. Beneficially, GMCs have higher yields than non-modified crops; farmers do not need to expand their acreage in order to produce more. On the other hand, with increased yields, farmers may choose to expand to produce even more. Such expansion has been observed in South American farms. Additional evidence suggests that genetically modified crops have reduced the populations of several different pests.

Sustainable Agriculture

A counter to industrial agricultural practices, sustainable agriculture is the practice of producing commodities for this generation without harming the resources for future generations. Common practices in sustainable agriculture include a reduction in the use of harmful chemicals, methods to promote good soil, and less water use. Sustainable agriculture also focuses on farmworker well-being and environmental conservation. Another component of sustainable agriculture is the production of organic crops, or those raised without synthetically processed herbicides. Using techniques such as crop rotation and natural pest management, organic crop production is meant to be an alternative to the concerns and criticisms surrounding conventional agriculture. Another facet of organic farming

Composting is a major component of sustainable agriculture.

is composting. The decomposition of organic, or living, materials returns nutrients to the soil. Proponents of organic and sustainable agriculture argue that these practices will ensure viable resources, like soil and ecosystems, for the future. The demand for sustainable agriculture and organic farming has also increased. As of 2013, 2.8 million hectares (6.9 million acres) of land in North America is used for organic agriculture.

Even biotechnology giants like Monsanto are getting in on the organic game. Part of their production includes plants that are not genetically modified; they use ancient crossbreeding hybridization techniques to ensure the best possible traits and entice buyers that might otherwise not purchase GMCs.

Today, there are approximately fourteen thousand certified organic farms in the United States, and over 186,000 acres (75,000 hectares) of farmland are used for certified organic agriculture. Organic farming profits around $5 billion each year. California, Wisconsin, New York, and Washington State have the highest number of organic farms.

The scientific journey to producing genetically modified crops is a long and complex one. From the ancient domestication of crops to Mendel's experiments, from the discovery of DNA to the first transgenic plant, researchers have a powerful yet controversial tool. Are genetically modified crops the key to ending global hunger? Will they ultimately improve agriculture's effects on the environment? Research will continue to provide the answers as the way toward new discoveries is paved.

Neil de Grasse Tyson on GMCs

One of the most well-known scientists today, astrophysicist Neil de Grasse Tyson, publicly stated his support for genetically modified organisms in a video posted to YouTube. His argument centers on the fact that humans have been modifying organisms for thousands of years.

Every food product, Tyson argues, contains some version of a genetically modified organism. He points to the fact that there are no wild cows and no naturally occurring seedless watermelons. If a food does have a wild component, Tyson points out, it is probably not as large or productive as the ones humans domesticated for consumption.

In June 2017, the documentary *Food Evolution* premiered. The documentary explores the controversy surrounding genetically modified crops. Specifically, the documentary raises many ethical questions: How can society make sure that there is enough food for everyone to eat? How do we find sustainable ways to feed the world? Does genetic engineering always rely on pesticide use?

Tyson narrates the film, which paints genetically modified organisms in a positive light. He states that "the film explores all the ways science has been used and abused in public discourse surrounding the genetic engineering of food."

Critics of Neil de Grasse Tyson's arguments point out that inserting DNA from a different organism is not the same as selective breeding practices enacted by humans since the Fertile Crescent. Others express concern that the documentary is biased toward large corporations like Monsanto that have expansive marketing budgets.

Tyson's viewpoints illustrate the tension often felt between the opposing sides of the genetic modification debate. On the one hand, critics worry about the safety of such organisms and their long-term effects on the environment. Supporters, like Tyson, contend that genetically modified crops are not so different than other agricultural practices and that their benefits strongly outweigh any concerns.

Chronology

12,000 BCE

Natufians gather wild grains.

8000 BCE

Squash is grown in the Americas.

3000 BCE

Sumerians document their invention of the plow.

2180 BCE

Historical evidence points to a famine in ancient Egypt that may have prompted the empire's downfall.

2000 BCE

Hybrid rice is bred in Japan.

1000 BCE

Farmers grow fruit using grafting.

1200 CE

Serfdom becomes a primary governing structure in medieval Europe.

1400 Serfdom declines.

1750 England's agricultural revolution begins.

1809 Charles Robert Darwin is born.

1845 A fungal blight wipes out the Irish potato crop, a staple of the poor in Ireland.

1859 Darwin publishes *On the Origin of the Species.*

1865 Gregor Mendel begins his pea plant experiments.

1875 Luther Burbank moves to California after the successful sale of his Burbank potato variety.

1879 William James Beal begins his buried seed bottle experiment.

1900 Mendel's work is rediscovered by de Vries, Correns, and Seysenegg.

1909 Thomas Morgan begins his experiments on fruit fly inheritance in his lab, nicknamed the "fly room."

1930 The dust bowl forces thousands of midwestern farmers to flee their farms.

1945 The pesticide DDT is available for public purchase.

1950 Edwin Chargaff determines that nitrogen bases are present in DNA in equal amounts.

1962 Marshall Nirenberg discovers the central dogma.

1963 Francis Crick and James Watson receive the Nobel Prize for their discovery of the structure of DNA.

1972 Paul Berg uses recombinant DNA to create a chimera organism.

1976 Monsanto releases Roundup.

1980 The Supreme Court decides that patents could be awarded to living things, like genetically modified crops.

1982 The first transgenic plant, tobacco, is created.

1983 Kary Mullis develops polymerase chain reaction, a fast method for creating DNA copies.

1984 The Flavr Savr tomato is released.

1985 Plant Genetic Systems announces that it has developed antibiotic-resistant tobacco plants.

1996 The first resistant weed, *Lolium rigidum*, a grass species, is discovered in an apple orchard in Australia.

2002 The European Union decides that all genetically modified crops must go through a premarket approval process and must follow strict labeling requirements.

2003 The Human Genome Project is completed.

2013 Fraley, Chilton, and Van Montagu receive the World Food Prize. The activist group March Against Monsanto holds its first protest march.

Glossary

biodiversity A variety of living things in a given ecosystem.

biofuel Fuel made from an organic source.

bud A growth on a plant that develops into a leaf, flower, or shoot.

cell The basic unit of all living things.

chromosome A structure found in the nucleus of cells that carries genetic information in the form of genes.

cloning vector A segment of DNA that can be used to insert foreign DNA into a cell.

crop rotation Growing different varieties of crops in a given growing season.

cross-pollination The pollination of a plant by another plant's pollen.

DNA Deoxyribonucleic acid, the molecule that contains genetic information for almost all living things.

domestication To adapt a plant or animal for use by humans.

double helix A structure made from two double-stranded molecules.

embryonic Describes the early stage of development of multicellular, eukaryotic organisms.

enzyme A biological molecule that speeds up chemical reactions in living organisms.

eukaryote An organism consisting of a cell or cells in which genetic information is stored as DNA in chromosomes in the nucleus.

evolution The process by which characteristics of organisms change over time.

fallow farming A type of farming where land is plowed and tilled but left unseeded during a growing season.

foraging Searching for food in the wild.

genes Units of heredity that are passed on to offspring from parents in pairs.

genetic Relating to heredity.

hectare A metric unit of square measure, equal to 2.471 acres or 10,000 square meters.

heredity The passing down of genetic information from one generation to the next.

horticulturalist An individual that grows plants.

hunter-gatherers A group of humans that harvest their food from the wild.

hybridization The process of breeding two organisms from different species.

intron A segment of DNA that does not code for proteins and can also stop the DNA replication process.

irrigation Delivering water to agricultural crops at scheduled intervals.

mRNA A type of nucleic acid that relays information from DNA to the part of the cell where proteins are made.

natural selection The process by which organisms with the best characteristics survive and reproduce.

nucleotide The basic structure of DNA composed of a phosphate, sugar, and nitrogen base pair.

nucleus A structure found within eukaryotic cells that contains genetic material and controls cellular processes.

ornithology The study of birds.

pathogens Bacteria or viruses that can cause disease in living things.

polymerase chain reaction A process that copies segments of DNA.

prokaryotes A single-celled organism that does not have a nucleus, such as bacteria.

recombinant DNA DNA formed from the combination of genetic information from different organisms.

self-fertilization The reproductive process by which plants utilize their own pollen for fertilization.

serf An agricultural laborer bound under the feudal system to work on his lord's estate.

shoot A young part of a plant, like a branch.

transgenic Relating to an organism that contains genetic material into which DNA from an unrelated organism has been artificially introduced.

vegetative reproduction A form of asexual reproduction in which offspring are grown from a piece of a plant.

Further Information

BOOKS

Hesterman, Oran B. *Fair Food: Growing a Healthy, Sustainable Food System for All*. New York: Public Affairs, 2012.

Jenkins, McKay. *Food Fight: GMOs and the Future of the American Diet*. New York: Avery, 2017.

Langwith, Jacqueline. *Cloning*. Farmington Hills, MI: Greenhaven Press, 2012.

Pence, Gregory E. *Designer Food: Mutant Harvest or Breadbasket of the World?* Lanham, MD: Rowman & Littlefield, 2002.

Ronald, Pamela C. *Tomorrow's Table: Organic Farming, Genetics, and the Future of Food*. Oxford, UK: Oxford University Press, 2018.

Standage, Tom. *An Edible History of Humanity*. New York: Walker & Company, 2010.

WEBSITES

Genetic Literacy Project
https://geneticliteracyproject.org

The GLP is a nonprofit organization that seeks to educate the public about biotechnology.

Growing a Nation: The Story of American Agriculture
https://www.agclassroom.org/gan/timeline/index.htm

This website provides a timeline of agricultural innovation in the United States.

Monsanto
https://monsanto.com

Monsanto's website provides information and updates on its products and research.

The United States Department of Agriculture
https://www.usda.gov

The USDA's website provides an overview of agriculture practices and policies in the United States.

Bibliography

"Beyond Silent Spring: An Alternate History of DDT." Chemical Heritage Foundation, February 14, 2017. https://www.chemheritage.org/distillations/magazine /beyond-silent-spring-an-alternate-history-of-ddt.

"Certified Organic Survey 2016 Summary." United States Department of Agriculture, September 2017. http://usda.mannlib.cornell.edu/usda/current/OrganicProduction/OrganicProduction-09-20-2017_correction.pdf.

Fedak, George. "Marquis Wheat." *Canadian Encyclopedia*, accessed February 4, 2018. https://www.thecanadianencyclopedia.ca/en/article/marquis-wheat.

Fraley, R., et al. "Expression of Bacterial Genes in Plant Cells." *Proceedings of the National Academy of Sciences of the United States of America USA* 80 (1983): 4803–4807. http://www.pnas.org/content/pnas/80/15/4803.full.pdf.

Garthwaite, Josie. "Beyond GMOs: The Rise of Synthetic Biology." *Atlantic*, September 25, 2014. https://www.theatlantic.com/technology/archive/2014/09/beyond-gmos-the-rise-of-synthetic-biology/380770/.

Gepts, P. "Domestication of Plants." *Encyclopedia of Agriculture and Food Systems* (2014): 474–486. doi:10.1016/b978-0-444-52512-3.00231-x.

Giaimo, Cara. "The World's Longest-Running Experiment Is Buried in a Secret Spot in Michigan." Atlas Obscura, January 30, 2017. https://www.atlasobscura.com/articles/the-worlds-longestrunning-experiment-is-buried-in-a-secret-spot-in-michigan.

Gilla, Carey. "Corrected Timeline: History of Monsanto Co." Reuters, November 11, 2009. https://www.reuters.com/article/us-food-monsanto/corrected-timeline-history-of-monsanto-co-idUSTRE5AA05Q20091111?irpc=932.

"Grain: World Markets and Trade." United States Department of Agriculture Foreign Agricultural Service, February 2018. https://apps.fas.usda.gov/psdonline/circulars/grain.pdf.

"Herbert W. Boyer and Stanley N. Cohen." Chemical Heritage Foundation, May 26, 2017. https://www.chemheritage.org/historical-profile/herbert-w-boyer-and-stanley-n-cohen.

Jaffe, Greg. "What You Need to Know About Genetically Engineered Food." *Atlantic*, February 7, 2013. https://www.theatlantic.com/health/archive/2013/02/what-you-need-to-know-about-genetically-engineered-food/272931/.

Lau, Jessica. "Same Science, Different Policies: Regulating Genetically Modified Foods in the United States and Europe." Science in the News, August 10, 2015. http://sitn.hms.harvard.edu/flash/2015/same-science-different-policies.

Lee, Richard B., and Richard Daly. *The Cambridge Encyclopedia of Hunters and Gatherers*. Cambridge, UK: Cambridge University Press, 2010.

"Luther Burbank." Luther Burbank Home and Gardens, accessed February 04, 2018. http://www.lutherburbank.org/about-us/luther-burbank.

Mudge, Ken, Jules Janick, Steven Scofield, and Eliezer E. Goldschmidt. "A History of Grafting." *Horticultural Reviews* 35 (2009): 437–493. doi:10.1002/9780470593776.ch9.

Neuman, William, and Andrew Pollack. "Farmers Cope with Roundup-resistant Weeds." *New York Times*, May 03, 2010. http://www.nytimes.com/2010/05/04/business/energy-environment/04weed.html?pagewanted=all.

Nicolia, Alessandro, Alberto Manzo, Fabio Veronesi, and Daniele Rosellini. "An Overview of the Last 10 Years of Genetically Engineered Crop Safety Research." *Critical Reviews in Biotechnology* 34 (2013): 77–88. doi:10.3109/07388551.2013.823595.

Paynter, Ben. "Monsanto Is Going Organic in a Quest for the Perfect Veggie." *Wired*, June 03, 2017. https://www.*wired*.com/2014/01/new-monsanto-vegetables.

Pray, Leslie. "Transposons, or Jumping Genes: Not Junk DNA?" Nature News, 2008. https://www.nature.com/scitable/topicpage/transposons-or-jumping-genes-not-junk-dna-1211.

Regis, Natalie. *Genetically Modified Crops and Food*. New York: Britannica Educational Publishing, 2016.

Richardson, Joel. "Mary-Dell Chilton: The Queen of *Agrobacterium*." Syngenta *Thrive*, 2016. http://www.syngenta-us.com/*thrive*/community/mary-dell-chilton.html.

Ronald, Pamela. "Plant Genetics, Sustainable Agriculture and Global Food Security." *Genetics* 188 (2011): 11–20. doi:10.1534/genetics.111.128553.

"The Rosalind Franklin Papers: Biographical Information." United States National Library of Medicine, November 17, 2015. https://profiles.nlm.nih.gov/ps/retrieve/Narrative/KR/p-nid/183.

Vaeck, Mark, et al. "Transgenic Plants Protected from Insect Attack." Nature News, July 2, 1987. https://www.nature.com/articles/328033a0.

Van Montagu, Marc. "The Irrational Fear of GM Food." *Wall Street Journal*, October 22, 2013. https://www.wsj.com/articles/the-irrational-fear-of-gm-food-1382481840.

"What Is Sustainable Agriculture?" University of California, Davis Sustainable Agriculture Research and Education Program, accessed February 19, 2018. http://asi.ucdavis.edu/programs/sarep/about/what-is-sustainable-agriculture.

"World Hunger Statistics." Food Aid Foundation, accessed January 21, 2018. http://www.foodaidfoundation.org/world-hunger-statistics.html.

Zimmermann, Kim Ann. "Pleistocene Epoch: Facts About the Last Ice Age." LiveScience, August 29, 2017. https://www.livescience.com/40311-pleistocene-epoch.html.

Index

Page numbers in **boldface** are illustrations.

About the Author

Megan Mitchell is a former biology educator. She works for a food literacy and school garden nonprofit. She has written many lesson plans for her high school biology classroom and designed middle school science curriculum. She received her MS in community development from the University of California, Davis. *Great Discoveries in Science: Genetically Modified Crops* is her fourth full-length book. She also wrote *The Human Genome* and *Jane Goodall: Primatologist and UN Messenger of Peace*. Mitchell loves gardening, hiking, reading, and performing scientific experiments. She hopes that one day genetics will be able to explain why her two dogs, Butters and Toast, have such silly personalities.